Developments in Heat Transfer

Developments in Heat Transfer

Editor

Petros Antonis

scitus
academics

Developments in Heat Transfer
Edited by **Petros Antonis**

ISBN: 978-1-68117-306-1
Library of Congress Control Number: 2016937614

SCITUS Academics LLC,
www.scitusacademics.com
Box No. 4766, 616 Corporate Way,
Suite 2, Valley Cottage,
NY 10989

Notice

Printed in United States of America on Acid Free Paper

Preface

Increased heat transfer is an important part of the heat transfer experiment and theory. Three methods can transfer heat from one location to other location: conduction in solids, convection of fluids (liquids or gases) and radiation through anything that allows radiation to pass. The heat transfer method is usually the most efficient. If a system has a difference in temperature, the heat will always move from higher to lower. Heat transfer between a high temperature object and a lower temperature object usually takes place. The heat flows to the cooler object from the warmer to the same temperature. At the point where both objects touch, the faster molecules of the warmer object collide with the slower moving molecules of the cooler object. When they collide, the faster molecules give some of their energy to the slower molecules. The slower molecules get more thermal energy in the cooler object and collide with other molecules. This process continues until the heat energy of the warmer object is spread across the cooler object. Convection is usually the most efficient method for heat transfer in liquids and gases. Convection occurs when warmer fluid or gas areas are placed in the fluid or gas cooler areas. This cycle leads to a continuous circulation pattern and heat is transferred to cooler regions. The heat transfer deliberation was structured around some daily examples, such as the cooling of a hot coffee mug and the warming of a cold pop can. In order to give an overview of the current state of this important technology, representative developments have been mentioned and commented on in each category of improvement techniques over the past decades.

This book focuses on experimental and theoretical research, and innovative techniques of heat transfer to be useful for many researchers, scientists, engineers and graduates.

Table of Contents

Chapter 1	**Heat Transfer and Pressure Drop Characteristics in Straight Microchannel of Printed Circuit Heat Exchangers 1**
Chapter 2	**Condensation Heat Transfer in Horizontal Non-Circular Microchannels 35**
Chapter 3	**FEM Approach for Transient Heat Transfer in Human Eye 65**
Chapter 4	**MEMS-Based Boiler Operation from Low Temperature Heat Transfer and Thermal Scavenging 87**
Chapter 5	**Influence of Orientation and Radiative Heat Transfer on Aluminum Foams in Buoyancy-Induced Convection 113**
Chapter 6	**Combined Effects of Pipe Diameter, Reynolds Number and Wall Heat Flux and on Flow, Heat Transfer and Second-Law Characteristics of Laminar-Transitional Micro-Pipe Flows 143**

Chapter 7 **Enhanced convective heat transfer using graphene dispersed nanofluids 187**

Chapter 8 **Exergy Analysis of a Ground-Coupled Heat Pump Heating System with Different Terminals 211**

Chapter 9 **Experimental investigation of copper-grooved micro heat pipes (MHPs) 235**

Chapter 10 **Heat Transfer Characteristics of Square Micro Pin Fins under Natural Convection 251**

Chapter 11 **CFD simulation of the effect of particle size on the nanofluids convective heat transfer in the developed region in a circular tub 275**

Chapter 12 **Calculation Method for Forced-Air Convection Cooling Heat Transfer Coefficient of Multiple Rows of Memory Cards 293**

Index 309

1

CHAPTER

HEAT TRANSFER AND PRESSURE DROP CHARACTERISTICS IN STRAIGHT MICROCHANNEL OF PRINTED CIRCUIT HEAT EXCHANGERS

Jang-Won Seo [1], Yoon-Ho Kim [2], Dongseon Kim [3], Young-Don Choi [1,*] and Kyu-Jung Lee [1]

[1] *Department of Mechanical Engineering, Korea University, Seoul 136-713, Korea*

[2] *Samsung Electronics Co., Ltd., Maetan 3-dong, Yeongtong-gu, Suwon 443-742, Korea; E-Mail*

[3] *Department of Mechanical Engineering, Korea National University of Transportation, Chungbuk 380-702, Korea*

ABSTRACT

Performance tests were carried out for a microchannel printed circuit heat exchanger (PCHE), which was fabricated with micro photo-etching and diffusion bonding technologies. The microchannel PCHE was tested for Reynolds numbers in the range of 100⍰850 varying the hot-side inlet temperature between 40 °C–50 °C while keeping the cold-side temperature fixed at 20 °C. It was found that the average heat transfer rate and heat transfer performance of the countercurrrent configuration were 6.8% and 10%⍰15% higher, respectively, than those of the parallel flow. The average heat transfer rate, heat transfer performance and pressure drop increased with increasing Reynolds number in all experiments. Increasing inlet temperature did not affect the heat transfer performance while it slightly decreased the pressure drop in the experimental range considered. Empirical correlations have been developed for the heat transfer coefficient and pressure drop factor as functions of the Reynolds number.

Keywords: microchannel; printed circuit heat exchanger (PCHE); micro photo-etching; diffusion bonding; counterflow

1. INTRODUCTION

A printed circuit heat exchanger (PCHE) is fabricated by diffusion-bonding thin metal plates which were previously engraved with flow channels using chemical erosion techniques. As illustrated in Figure 1, the PCHE appears solid and seamless as the diffusion-bonding technique promotes growth of crystal grains between metallic surfaces that are compressed at a high temperature below the melting point [1], allowing for light weight and high structural strength. The micro photo-etching technique—which has been developed with the progress of MEMS technology—enables the processing of sophisticated microchannels on the metal surface. Additionally, development in the MEMS field has also allowed for easy mass production, reliability and economic efficiency [2,3].

Generally, for a conventional heat exchanger, the brazing technique—where bonding occurs by melting a binder—is widely used. A microchannel PCHE created through diffusion bonding has superior heat resistance and bonding strength than one created by the brazing technique. Because there is almost no thermal resistance, nor reduction or clogging of the microchannels at the time of bonding, excellent production properties and thermal performance can be achieved. Because of these advantages, it is possible to use this microchannel PCHE—created through diffusion bonding—in various fields such as fuel cell systems, chemical reaction processes, marine and terrestrial plants, and refrigeration and air conditioning systems, and the potential fields of use continue to expand [4,5].

Figure 1. Flow cross-section of a printed circuit heat exchanger (PCHE) fabricated using diffusion bonding [1].

Among the previous studies on micro heat exchangers, Peng et al. [6] conducted a study on the effect of the convection heat transfer coefficient on laminar and turbulent flows by using a rectangular microchannel. They determined that the degree of influence on the convective heat transfer coefficient is different, but the hydraulic diameter of each channel, and the gap between aspect ratio and channel under laminar and turbulence flow are important factors. Lee et al. [7] studied the local convective heat transfer characteristics of the rectangular microchannel through a numerical analysis. They found that as the Reynolds number is increased, the heat transfer performance was improved. Also, through comparison of numerical analysis and experimental results, Qu et al. [8]

concluded that there is no difference in the macro-sized channel heat exchanger in terms of the flow in the rectangular microchannel. Shen et al. [9] conducted a study on Poiseuille number, local Nusselt number and the surface roughness in a rectangular microchannel. They reported that the friction factor in the microchannel of laminar flow was measured larger than predicted, and the local and average Nusselt number was smaller than the predicted value. In addition, they suggested experimental correlations as functions of the Reynolds number for friction factor and the Nusselt number. Rachkovskij et al. [10] conducted a study on a cross-flow heat exchanger with laminated layers of 20 sheets and the aspect ratio of the microchannel, and in this experiment, air-air was used as the working fluid. From their research results, they suggested an optimal temperature proximity and volume heat transfer coefficient. Kang et al. [11] suggested a theoretical model which can be used to predict the heat and fluid properties of a micro-cross-flow heat exchanger. Nikitin et al. [5] experimentally investigated the heat transfer and pressure drop characteristics of supercritical CO2. They also proposed empirical correlations for the local heat transfer coefficient and the pressure drop factor as functions of the Reynolds number. Ngo et al. [12] have manufactured a new PCHE with an s-shaped pin by improving Nikitin et al.'s [5] straight channel PCHE and conducted experiments on this new PCHE. In addition, they evaluated the thermal hydraulic performance through a numerical analysis. Tsuzuki et al. [13] performed a numerical analysis for s-shaped and various zigzag-shaped PCHEs. They evaluated the thermal hydraulic performance by calculating the heat transfer and pressure drop. Ma et al. [14] performed a numerical analysis for the offset bubble and the offset-strip fin configurations. In this study, a cross-flow pattern was employed for the offset-bubble configuration pattern while the offset-strip fin's flow direction involved dispersion of the working fluid. Ma et al.'s [14] numerical analysis model focused on one region of the repeated small channels, using symmetric and periodic boundary conditions. The numerical analysis method assumed an incompressible fluid and used a governing equation while using Nusselt number and friction factor to analyze and compare results for both configurations. Ma et al. [15] manufactured a PCHE—through a photo-chemical etching method—with thermal plates of an airfoil channel configuration. This PCHE was then analyzed using a numerical analysis method, followed by a grid test which corresponded with experimental data. In order to analyze the effect of the fin-endwall fillet, Ma et al. [15] then varied the pitch of the airfoil fins and analyzed the Nusselt number and friction factor results. Ma et al [16] performed a numerical analysis for a PCHE with a zigzag channel configuration. The pitch, length, and angle of the zigzag channels were varied and heat transfer characteristics for a laminar flow region of a Reynolds number range of 400–2000 were studied. This numerical analysis assumed a working fluid of air and helium with an inlet temperature of 900 °C. The numerical method results were analyzed using Nusselt number, Colburn j-factor and Fanning friction factor (f-factor). Baek et al. [17] investigated flow maldistribution and axial conduction in regards to PCHE header configuration. Using a numerical analysis method, the flow direction of the working fluid for both vertical and horizontal configurations was considered. Through the NTU method, the effectiveness was obtained and a Nusselt number correlation for microchannels was proposed. Bartel et al. [18] studied PCHEs within advanced nuclear reactors. Within these PCHEs, wavy channel and offset strip fin configurations were compared and analyzed. Furthermore, a Colburn j-factor and fanning friction factor (f-factor) was proposed. Figley et al. [19] researched PCHEs utilized in reactors with high-temperature regions. Using a flow analysis program—Fluent Software—numerical results were acquired. Correlation of the pressure drop results was compared with the numerical analysis, allowing for the validity to be confirmed. Through comparison of the mass flow rate and NTU, the performance effectiveness was calculated. Kim et al. [20], after manufacturing a PCHE heat transfer plate and creating a 3D model of this plate, performed a numerical analysis. By changing the geometric parameters of the fin arrangement, the pressure drop and heat transfer characteristics were investigated in regards to geometric properties. Through these heat transfer characteristics results, the Colburn j-factor, Nusselt number and Euler number were expressed. Kim et al. [20] also compared the Fanning friction factor in accordance with Reynolds number. Koo et al. [21] investigated the flow characteristics of a PCHE inlet through a 3-D Reynolds-averaged Navier-Stokes analysis. Two other surrogate models—the Krigin and radial basis neural networks—were also employed. Additionally, in accordance to the flow rate increase and channel number, the flow characteristics were compared and analyzed. Mylavarapu et al. [22] conducted a numerical analysis based on a model of a PCHE for high-temperature gas-cooled reactors. Using a Reynolds number region of less than 2300, existing formulas were compared with the proposed correlation and analyzed, with the results of the cold and hot sides considered independently of each other. According to the Reynolds number, the Fanning friction number and Nusselt numbers were calculated, and experimental data was compared with the circular pipe correlation. Xu et al. [23] conducted a study on the optimization of fin arrangement and channel configuration for PCHEs using supercritical CO2 as a working fluid. The fin dimensions were varied, involving an airfoil fin type and differing fin thickness, length, and width. The average Nusselt number and pressure drop results were analyzed, in accordance to the increasing Reynolds number. Yoon et al. [24] analyzed four PCHE configurations; straight, zigzag, s-shape and airfoil channel. A numerical analysis method was used, employing a 3D model of the minimum unit structure which removed the need for numerical construction of the entire PCHE. This allowed for the hydraulic diameter, Nusselt number and pressure drop to be compared. The two working fluids used were helium and CO2, with the Fanning for each of the two working fluids calculated and compared. In addition, as part of a cost analysis, the total cost of each of the three different channel configurations was calculated. Yoon et al. [25] conducted a study focusing on crossflow PCHEs within advanced small modular reactors. After confirming a design model, the MATLAB program was used to analyze through mathematical methods. First, a single-pass crossflow was designed and then partial differential equations were obtained by employing the Laplace transform and inverse transform. This allowed for solutions for each variable to be obtained. Yoon et al. [25] then calculated results for the thermal design process, cost estimation methodology, effectiveness and crossflow PCHE analysis. Jeong et al. [26] proposed enhancements to the plate fin type heat exchanger after modelling a fin type and louver fin heat exchanger configuration. Subsequently, in order to evaluate the grid reliability, friction factor and convergence grid tests were performed. The effective area factor was determined through calculation of the non-dimensional factor, Colburn j-factor, and Fanning friction factor (f-factor), and this performance of the commercial-fin configuration was compared with the proposed enhanced fin configuration. Kim et al. [27–29] conducted a numerical analysis for PCHEs with wavy channels of variable angles and with a hot-side double-banking heat plate arrangement. Kim et al. [27–29] proposed a heat transfer and pressure drop correlation for a working fluid of helium and a Reynolds number of 3000 or lower. Furthermore, Kim et al. [27–29] considered the cost of the system power loss, in regards to the stacked thermal plate layers, and analyzed the results to propose an improved PCHE design method.

Aside from these previous studies, research on PCHEs is rather limited, especially when considering the significant amount of research that has been conducted on other types of commercial heat exchangers. Furthermore, within the body of heat exchanger research there are few studies examining microfluidics and pressure drop characteristics and with most employing a Reynolds number less than 1000 in conjunction with an average and unchanging Nusselt number.

In this study, the authors have fabricated PCHE heat exchangers with straight-tube-shaped microchannels and obtained heat transfer and pressure drop data by varying the Reynolds number and the operating temperature. From the results, empirical correlations have been proposed for the heat transfer coefficient and friction factor, which can be used as the basic data for heat exchanger design.

2. EXPERIMENTAL SETUP AND DATA

2.1. Microchannel PCHE

The microchannels were formed using photo-etching technology on the cold and hot sides of the heat transfer plates, as shown in Figure 2. Each channel consists of an inlet, a straight middle and an outlet section, all having a half-moon shaped cross section. Two types of heat exchangers were fabricated with different structures. One (PCHE#1) has three hot-side plates and four cold-side plates, and the other (PCHE#2) has five hot-side and six cold-side plates, each with the hot and cold-side plates alternately layered. On the top and bottom of the layered heat transfer plates, extra (end) plates were bonded in order to increase structural strength. The structure and flow configuration are shown in Figure 3. The flow configuration was set for a counterflow to obtain a smaller approach temperature. Once the heat transfer plates were bonded, inlet and outlet ports were attached using electric welding. Due to a lack of gasket and the close distance between the hot fluid and cold fluid, the manufactured PCHEs have a high heat transfer rate.

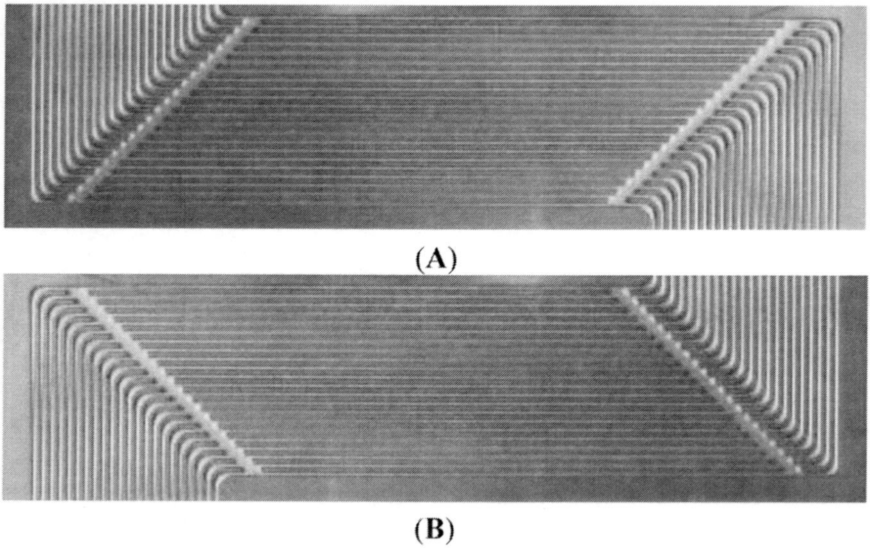

Figure 2. Photos of the metal-plates with straight middle sections. (A) Hot-side plate; (B) Cold-side plate.

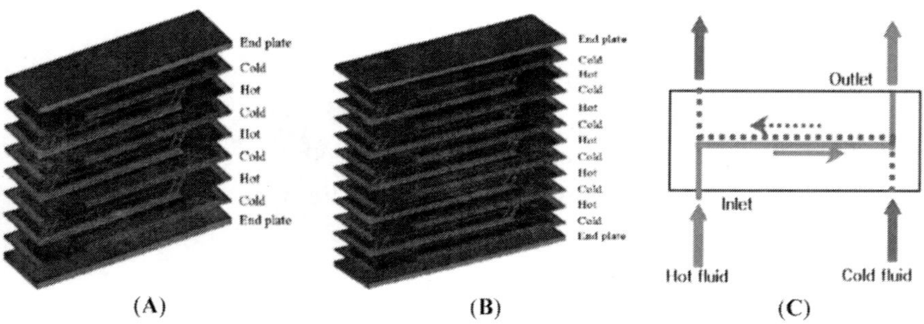

Figure 3. The stack layer and the flow pattern in the microchannel printed circuit heat exchanger (PCHE). (A) PCHE#1 (3 hot/4 cold); (B) PCHE#2 (5 hot/6 cold); (C) Flow configuration.

Figure 4 shows the microchannel PCHE used, and detailed specifications are listed in Table 1. One-quarter of the PCHE was cut out in order to confirm the shape of the internal channels and the bonding condition of heat transfer plates. The cut PCHE and the cross-sectional pictures of channel are shown in Figure 5. The cross-section shows half-moon shaped channels, characteristic of the micro photo-etching process employed. The entrance area (Ac) and the effective heat transfer (As) area were calculated considering the half-moon profile. Furthermore, as shown in the figure, the bonding conditions of the plates were excellent.

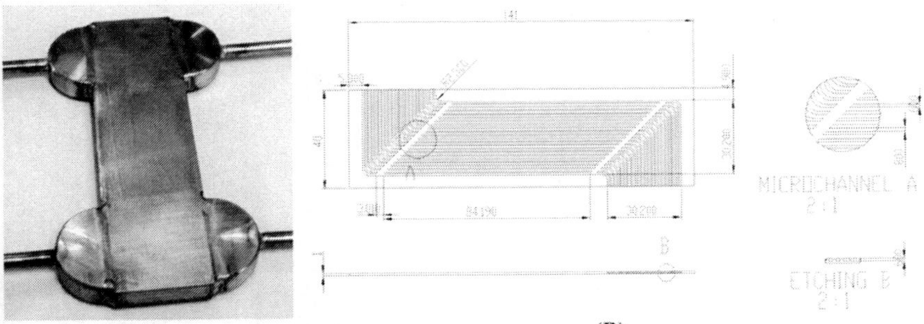

(A) (B)

Figure 4. The final shape of the microchannel printed circuit heat exchanger (PCHE). (A) The final shape of the PCHE; (B) Detail design drawing sheet.

Table 1. Microchannel printed circuit heat exchanger (PCHE) Specifications.

Metal-plate material		SUS304L
Dimensions of PCHE (W × L × H), mm		141 × 40 × 16
Dimensions of plates (W × L × H), mm		141 × 40 × 1
Dimensions of end plates (W × L × H), mm		141
Number of plates	Hot side	3, 5
	Cold side	4, 6
Number of channels per plate		22
Channel width		800 μm
Land (solid) width		600 μm
Channel height		600 μm

Figure 5. Cross-sectional view of a microchannel printed circuit heat exchanger (PCHE) fabricated through the diffusion-bonding method.

2.2. Experimental Setup

Figure 6 shows the experim ental setup. It consists of two sections, one circulating the hot fluid and the other circulating the cold fluid. In order to maintain constant inlet temperature and flow rate, each

(A)

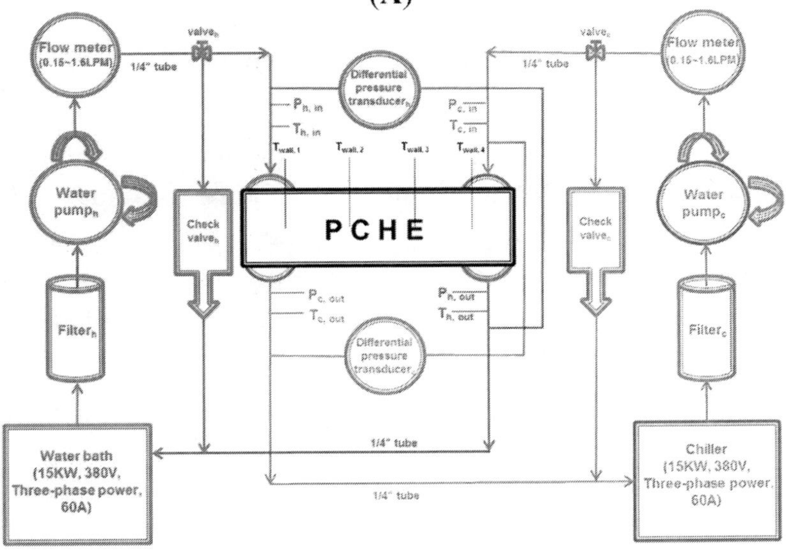

(B)

Figure 6. Schematic diagram and photograph of the experimental setup. (A) Photograph of the experimental setup; (B) Flow diagram of the experimental setup.

section has a thermostatic bath, a controllable magnetic gear pump, and a volumetric flowmeter. A filter was installed at the inlet of each flowmeter to remove foreign matter in the fluid and to prevent fluctuations in, and rusting of, the flow meter. Insulation has been provided all across the sections in the experimental setup in order to minimize heat loss. Thermocouples, as well as absolute and differential pressure gauges, were installed at all inlets and outlets. Prior to performing experiments, each measuring device was calibrated. Thereafter, the data of flow rate, temperature, pressure, differential pressure etc. were stored on a computer by using a data acquisition unit (DAQ). After the experimental setup had reached a pre-designated steady-state operating condition, all the measurements were stored at 5 s intervals.

2.3. Experimental Conditions and Results Analysis

Water was used as the hot and cold fluid. The inlet temperatures for the hot fluid were 40 °C and 50 °C. The experiment was performed while keeping the cold fluid's inlet temperature constant at 20 °C. The hot and cold flow rates were measured in a range of 0.377⍰1.391 L/min, where the flow and pressure drop were both stable. The Reynolds number was calculated in a range from 100⍰850.

The hydraulic diameter and Reynolds number are calculated using the method suggested by Cowell [30] as:

$$D_h = \frac{4A_c L_f}{A_s} \tag{1}$$

$$\mathrm{Re}_h = \frac{\rho V D_h}{\mu} = \frac{\dot{m} D_h}{\mu A_c} \tag{2}$$

where Ac is the free flow area, As is the total heat transfer area and Lf is the length of the flow stream in a channel. On the hot side, Ac is 31.7 mm2 and As is 26,037 mm2. On the cold side, Ac is 42.2 mm2 and As is 34,716 mm2. Lf is 137 mm and Dh is 0.6685 mm on both sides.

The heat transfer rate in the hot and cold fluids passing through the test section can be obtained using Equations (3) and (4):

$$Q_h = \dot{m}_h C_{p,h}(T_{h,i} - T_{h,o}) \tag{3}$$

$$Q_c = \dot{m}_c C_{p,c}(T_{c,o} - T_{c,i}) \tag{4}$$

The heat performance (UA) value can be obtained by using the logarithmic mean temperature difference (LMTD) and the average heat transfer rate, as represented by Equation (7):

$$\Delta T_{LMTD} = \frac{(T_{h,i} - T_{c,o}) - (T_{h,o} - T_{c,i})}{1\text{n}\left[\frac{(T_{h,i} - T_{c,o})}{(T_{h,o} - T_{c,i})}\right]} \tag{5}$$

$$Q_m = \frac{Q_h + Q_c}{2} \tag{6}$$

$$UA = \frac{Q_m}{\Delta T_{LMTD}} \tag{7}$$

Generally, the overall heat transfer coefficient can be calculated from the sum of the thermal resistances as in:

$$\frac{1}{UA} = \frac{1}{h_h A_h} + \frac{1}{h_c A_c} + \frac{t}{k A_m} \tag{8}$$

where t, k, and Am denotes the gap between the hot and cold side channels—which is 0.4 mm—the thermal conductivity of the heat transfer plate is 16.2 W/m·K, and the average heat transfer area respectively. The hot-side heat transfer coefficient, hh and the cold-side heat transfer coefficient, hc were obtained by using the modified

Wilson plot method [31]. The measurement error was calculated using Equation (9):

$$Q_{loss}(\%) = \frac{|Q_h - Q_c|}{Q_h} \tag{9}$$

Only the results within 7% error boundaries were selected as shown in Figure 7. The total pressure drop of the microchannel PCHE may be expressed as:

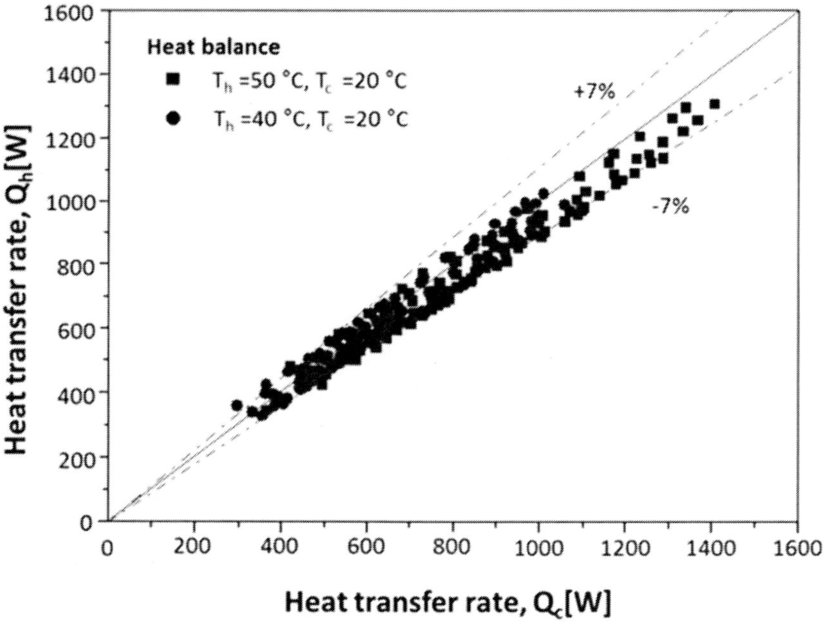

Figure 7. Heat balance between hot and cold sides.

$$\Delta P = \frac{1.5G_p^2}{2\rho_i} + \frac{4fLG^2}{2D_h}\left(\frac{1}{\rho}\right)_m$$
$$G_p = \frac{4\dot{m}}{\pi D_p^2} \tag{10}$$

where (1/ρ)m is the average density across the flow path and Gp denotes the mass flux at the inlet port. Note that the effect of hydrostatic pressure is neglected. The pressure drop was the measured sum of the microchannel, the inlet ports, and the outlet ports [32]. Experimental uncertainty was calculated by using ASME PEC 19.1 [33] and NIST Technical Note 1297 [34]. The total uncertainty consists of bias error and precision error as shown in Equation (11). When propagating errors, Equation (12) gives the uncertainty of the calculated parameters based upon the measured variables:

$$\Pi = 2\sqrt{\left(\frac{B}{2}\right)^2 + \left(\frac{S}{\sqrt{N}}\right)^2} \tag{11}$$

$$\Pi_p = \sqrt{\sum_{i=1}^{n}\left(\frac{\partial p}{\partial S_i}u_{si}\right)^2} \tag{12}$$

In Equations (11) and (12), Π is the total uncertainty, B is Bias error, S is a standard deviation, N is the number of measurements, and p is the computational variable. The experiments were conducted by repeating each measurement three times (N = 3). The detailed results for the uncertainty analysis in this experiment are presented in Table 2.

Table 2. Parameters and estimated uncertainty.

Parameters	Uncertainty (%)
Temperature, T	0.6
Pressure drop, ΔP	0.92
Flow rate of hot side, $\dot{m}_h$	1.19
Flow rate of cold side, $\dot{m}_c$	0.94
Averaged heat transfer rate, Q_m	1.19
Reynolds number of hot side	3.13
Reynolds number of cold side	3.29
Heat transfer coefficient of hot side	7.36
Heat transfer coefficient of cold side	7.31
Friction factor, f	5.8

3. EXPERIMENTAL RESULTS AND DISCUSSION

3.1. Heat Transfer Characteristics

Figures 8–10 show the heat transfer characteristics of the two types of heat transfer exchangers, i.e., PCHE#1 and PCHE#2 in Figure 3, measured for various Reynolds number conditions on the hot and cold sides. Figure 8A shows the average heat transfer rates measured keeping the same Reynolds numbers on both sides. Note that the flow rate of the cold fluid is larger than that of the hot fluid to maintain the same Reynolds number as there is one more plate (channel) on the cold side in the PCHEs (see Figure 3). Two cases of hot-side inlet temperature, i.e., 40 and 50 °C, were measured while cold-side inlet temperature was fixed at 20 °C. The heat transfer rate is larger for PCHE#2 than PCHE#1 as the same hot-side inlet condition and the influence of hot-side inlet temperature is also larger for PCHE#2.

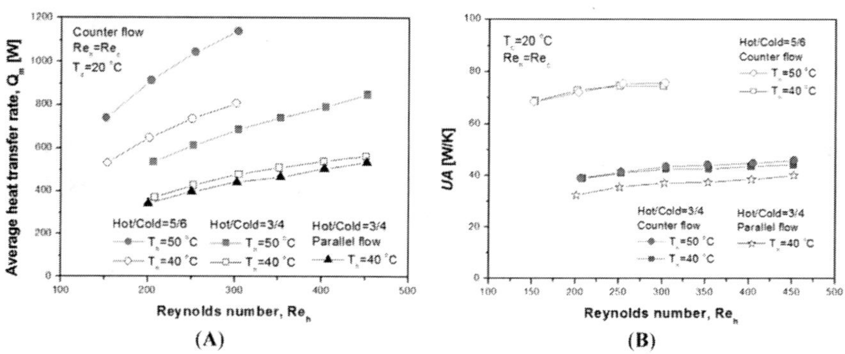

Figure 8. Average heat transfer rate and the heat performance (UA) with the same Reynolds number on hot and cold sides. (A) Average heat transfer rate vs. Reynolds number; (B) UA vs. Reynolds number.

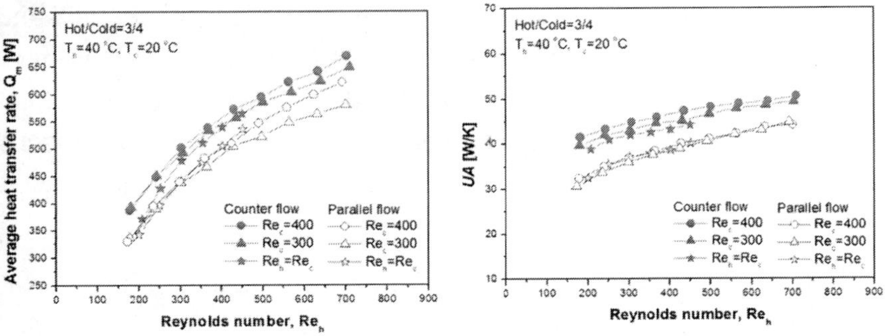

(A) (B)

Figure 9. Influence of flow configuration (countercurrent vs. parallel). (A) Average heat transfer rate vs. Reynolds number; (B) Heat performance (UA) vs. Reynolds number.

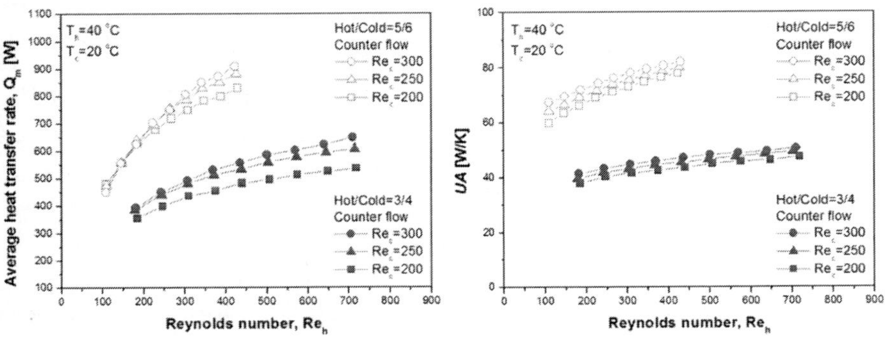

(A) (B)

Figure 10. Influence of stacked lamination (PCHE#1 vs. PCHE#2). (A) Average heat transfer rate vs. Reynolds number; (B) Heat performance (UA) vs. Reynolds number.

Figure 8B shows the corresponding UA values. The figure shows that the UA increases very slowly with an increasing Reynolds number and the UA value of PCHE#2 is larger—by about 1.8 times—than that of PCHE#1, which is expected from Figure 8A. It seems that the influence of the inlet temperature is minimal as the UA is practically the same for the different inlet conditions considered. In comparison with PCHE#1, PCHE#2 has more plates, having a larger free flow and heat transfer area. Since the flow conditions, i.e., Reynolds numbers, are the same, it may be assumed that U is the same in both heat exchangers. Then, the difference between the two heat exchangers' UAs in Figure 8A is attributable to the difference in heat transfer area.

In order to see the influence of flow direction, PCHE#1 was additionally tested in a parallel configuration for 40 °C hot-side inlet condition. The results were compared with the corresponding countercurrent results in Figure 9. The average heat transfer rate of countercurrent configuration is about 6.8% larger than that of parallel flow. The UA of the countercurrent flow is approximately 10%–15% larger than parallel flow. In the case of the countercurrent flow, logarithmic mean temperature difference is smaller than the parallel flow by ca. 0.3–1.0 K.

The same heat exchangers were measured again, varying only the hot-side Reynolds number while fixing the cold-side Reynolds number at 200, 250, and 300. The hot- and cold-side inlet temperatures were fixed at 40 °C and 20 °C, respectively. Figure 10 shows the results.

The experiment shows the tendency that the more the Reynolds number of the hot and cold sides increases, the more the average heat transfer rate and heat transfer performance increases. However, as the Reynolds number of the hot side is increased, the increase range in heat transfer rate and heat transfer performance narrows. This narrowing is the change of the hydrodynamic entry region as it becomes fully developed due to the increase in the mass flow rate of the hot side. Figure 10A shows that the Reynolds number of the cold side was 200, 250, and 300, and that the average heat transfer rate in PCHE#2 was 1.5 times more than that of PCHE#1. The heat transfer performance of PCHE#2, as indicated in Figure 10B, was approximately 1.6 times higher than PCHE#1.

In order to obtain a new heat transfer coefficient from a single-phase experiment, the UA value that previously represented the heat transfer performance was used to determine the correlation of heat transfer coefficient in this experiment. The Wilson plot method [35] is known as the method which, after calculating the overall heat transfer coefficient from a heat exchanger, obtains each heat transfer coefficient of the hot and cold sides using those values. Currently the modified Wilson plot method, used in various experimental ranges, is widely used. In this study, the modified Wilson plot method was used in order to obtain the heat transfer coefficient of the hot and cold sides, respectively. The heat transfer coefficient can be expressed as Equation (13) for the heat resistance balance of the hot and cold sides by using Re and Pr of power-law form:

$$\left[\frac{1}{UA}-\left(\frac{1}{kA_m}\right)\right]\left[\frac{k}{D_h}\mathrm{Re}^a\mathrm{Pr}^{1/3}A\right]_h=\frac{1}{C_h^{'}}+\frac{1}{C_c^{'}}\left[\frac{k}{D_h}\mathrm{Re}^a\mathrm{Pr}^{1/3}A\right]_h\left[\frac{k}{D_h}\mathrm{Re}^a\mathrm{Pr}^{1/3}A\right]_c^{-1} \tag{13}$$

where coefficients $C'h$ and $C'c$ and Re index were calculated using iterative multiple linear-regression analysis [36].

Figure 11 shows the results of the heat transfer experiments of the cold side, obtained by using the modified Wilson plot method. N represents the number of lamination layers which are the cold side. The convective heat transfer coefficient correlation of the cold side in the Reynolds number range is the same as Equation (14):

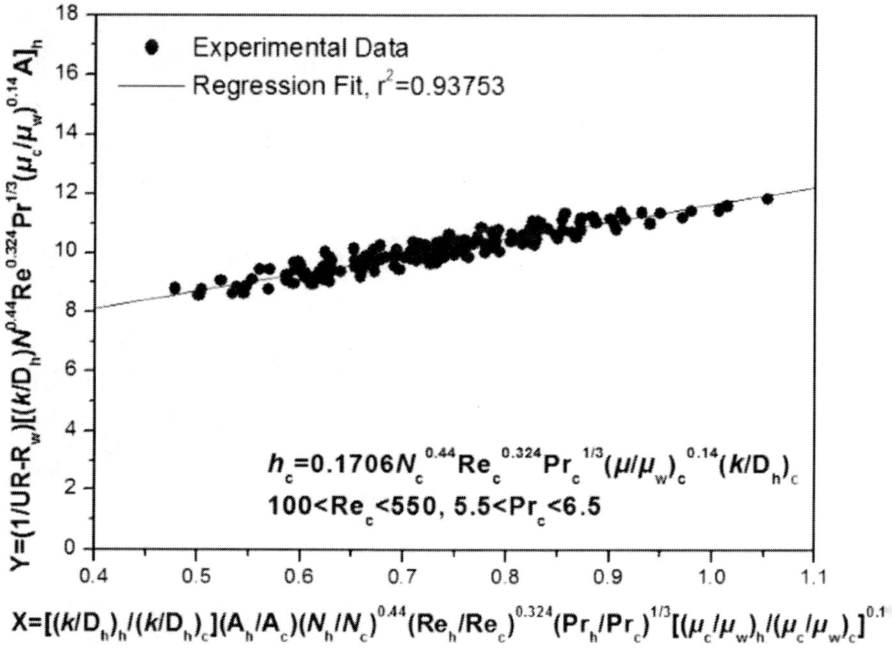

Figure 11. Typical modified Wilson plot results for the calibration of the cold-side heat transfer coefficient.

h c =0.1706N 0.44 c Re 0.324 c Pr 1/3 (μ c /μ w) 0.14 (k/D h) c ,100 <Re c <550

(14)

The heat transfer coefficient of the hot side can be obtained by using the heat transfer coefficient correlation proposed for the cold side. N indicates the number of lamination layers of the hot side. Equation (15) expresses the proposed convective heat transfer coefficient correlation:

$$h_h = 0.1729 N_h^{0.44} \mathrm{Re}_h^{0.324} \mathrm{Pr}^{1/3} (\mu_h/\mu_w)^{0.14} (k/D_h)_h \, , 100 < \mathrm{Re}_h < 850$$

(15)

Figure 12A shows an error range within 7% when comparing the experiment's results and Equation (15). Figure 12B shows the results of comparing the difference between the Nusselt number, which is non-dimensional form, and the proposed correlation (Equation (15)) by using the calculated convective heat transfer coefficient. If these results are represented in the form of a new correlation, including the variable of the number of lamination layers, it can be expressed as Equation (16):

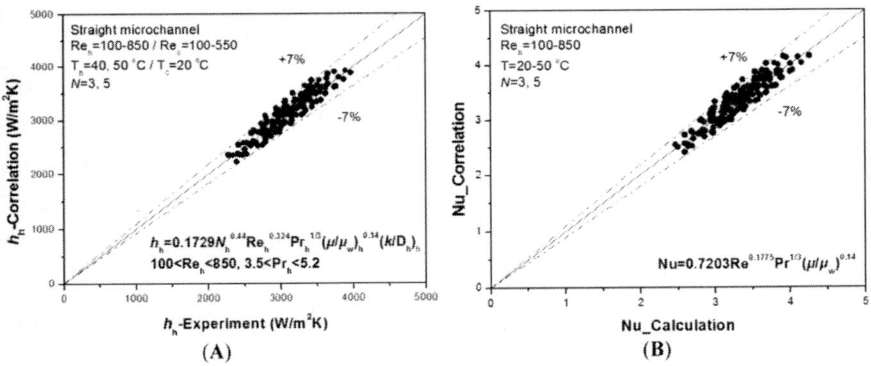

Figure 12. Comparision of suggested correlations and experimental data for hot-side heat transfer coefficients. (A) Heat transfer coefficient; (B) Nusselt number.

$$Nu_h = 0.7203\mathrm{Re}_h^{0.1775}\mathrm{Pr}^{1/3}\left(\mu_h/\mu_w\right)^{0.14} \tag{16}$$

The accuracy of the correlation within the range of 7% and the Reynolds number range from 100–850 were proposed for the range of this experiment.

3.2. Pressure Drop Characteristics

Figure 13 shows the pressure drop according to the change of the Reynolds number and temperature of the hot and cold sides. As the Reynolds number increases, the figure shows the tendency of the pressure drop to also increase. The increase of the Reynolds number represents the increase of the mass flow rate in the microchannel. This increase causes an increase in flow resistance, and as a result, the pressure drop will also increase. When the temperature of the hot and cold sides are 40 °C and 20 °C, respectively, the pressure drop, according to the change of the cold fluid's Reynolds number, shows that the Reynolds number for the hot and cold fluids increases equally as the pressure drop increases. In the range of this experiment, the pressure drop of the hot side, according to the change of the mass flow rate of the cold side, is not significantly influenced. If the inlet temperature of the hot side is increased to 50 °C, Figure 14A shows a slight pressure drop. This change results from the influence of viscosity and density according to the change of the inlet temperature of the hot fluid. As the inlet temperature increases, density and viscosity are reduced. On the other hand, the UA indicating the heat transfer performance shows almost the same performance and is not affected by the inlet temperature.

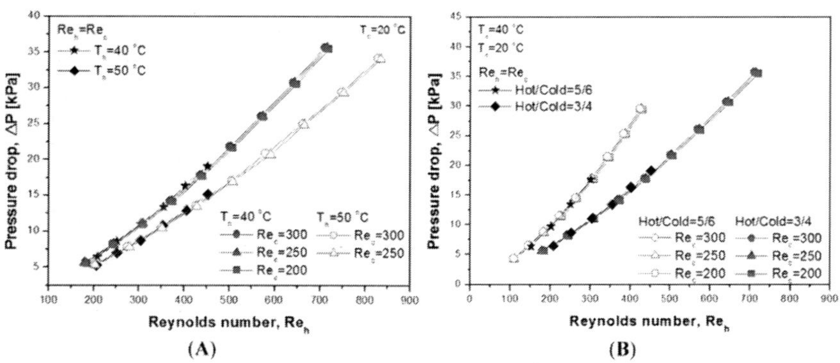

Figure 13. Pressure drop vs. Reynolds number in all experiments. (A) Difference in inlet temperature; (B) Difference in number of lamination layers.

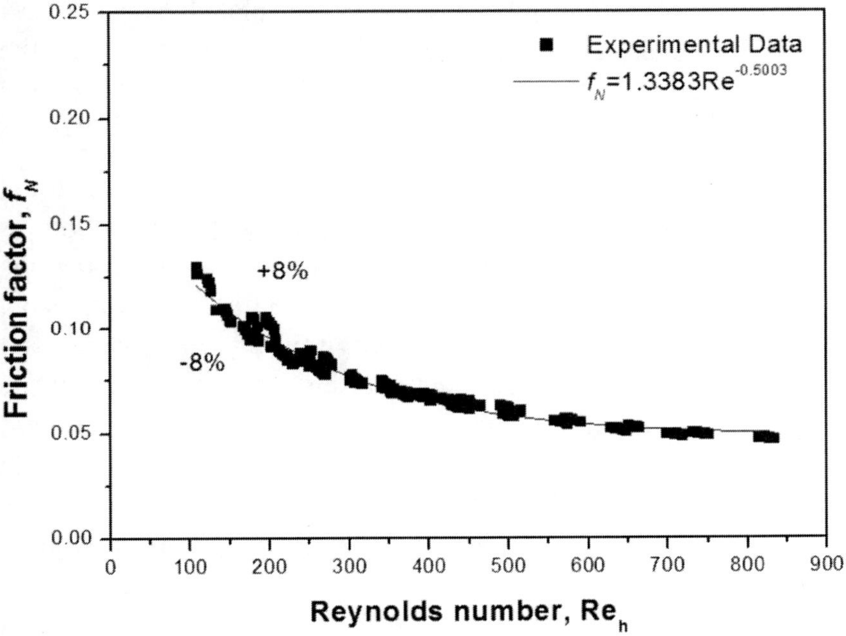

Figure 14. Comparison of the friction factor correlation and experimental data for the microchannel printed circuit heat exchanger (PCHE).

Figure 14 shows a graph using Equation (10), which is the theoretical equation that represents the pressure drop. Equation (16) expresses the friction factor fN, which is the result value of the pressure drop according to Reynolds number. N indicates the number of lamination layers of the hot side. The total pressure drop was divided by the number of lamination layers. The friction factor correlation is represented by the function of the Reynolds number, and is as follows:

$$f_N = 1.3383\mathrm{Re}^{-0.5003}, 100 < \mathrm{Re} < 850 \tag{17}$$

The exponent of the Reynolds number was calculated using the least squares method. The accuracy of the correlation and the experimental results are within ±8%, and the coverage of the Reynolds number is 100–850.

Figure 15 shows comparison of the proposed non-dimensional form, Colburn j-factor, and friction factor. The proposed correlation was compared with the Kays and London correlation [37] for the corrugated surface and offset strip fin configuration, which utilized a Reynolds number range of 400–3000. Within the flow direction of these two improved channel configurations, the working fluid is disturbed, resulting in a higher heat performance than the straight channel configuration. It was found that the offset-strip-fin configuration had the highest heat performance, followed by the corrugated-surface configuration. Inversely, the straight microchannel exhibited the lowest friction factor, followed by the corrugated surface and with the offset strip fin configuration having the highest friction factor.

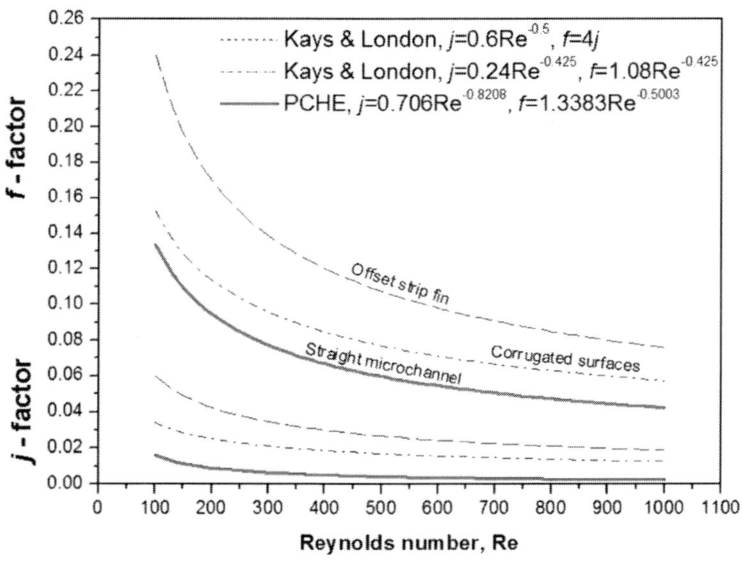

Figure 15. Comparison of the proposed j-factor and f-factor correlation with previous correlations.

4. CONCLUSIONS

In this study, the single-phase experiment for the characteristics of the heat transfer and the pressure drop of the microchannel PCHE was carried out. Based on this single-phase experiment, the following values and characteristics of the heat transfer coefficient and friction coefficient correlation were proposed.

The average heat transfer rate of the counterflow PCHE is about 6.8, and the UA of the heat transfer performance is excellent to the extent of approximately 10%–15%.

As the Reynolds number of the hot and cold sides increases and the inlet temperature increases, the average heat transfer rate also increases. This increase was the general performance characteristic of the heat exchanger according to the increase of the flow rate.

As the Reynolds number of the hot and cold sides increases, the pressure drop increases. If the inlet temperature of the hot side is constant, the pressure drop according to the change of Reynolds number of the cold side shows equal results.

The heat transfer performance is not affected by the change in the inlet temperature of the hot side, but if the inlet temperature is high at the time of the pressure drop, which shows a slight pressure drop.

The heat transfer coefficient correlations of the hot and cold sides using the modified Wilson plot method are proposed. The Reynolds number range of these correlations is 100–850.

The friction factor fN was calculated using the pressure drop results. The application scope is the same as above. It is expected that the experimental results obtained in this study will be usable as the basis for future performance experimental data.

ACKNOWLEDGMENTS

The authors would like to acknowledge the late Dr. Kyu-Jung Lee (Korea University, College of Mechanical Engineering) for his mentorship and for sharing his expertise in heat transfer.

AUTHOR CONTRIBUTIONS

The initial research concept and research design was completed by Jang-Won Seo, Yoon-Ho Kim and Kyu-Jung Lee. Jang-Won Seo conducted background research into previous studies. Yoon-Ho Kim manufactured the PCHE used in the study. Jang-Won Seo constructed the experimental apparatus and performed all experiments. Jang-Won Seo and Dongseon Kim undertook data analysis. This manuscript was primarily written by Jang-Won Seo, with Dongseon Kim and Young-Don Choi providing assistance. All authors have read and approved the final manuscript.

Nomenclature

Ac	Minimum free flow area (mm)2
As	Total effective heat transfer area (mm)2
B	Bias error
Cp	Specific heat (J/kg·K)
Dh	Hydraulic diameter (mm)
f	Friction factor

G	Core mass velocity (kg/m2·s)
Gp	Fluid mass velocity in the port (kg/m2·s)
H	Thickness of metal sheet (mm)
j	Colburn j-factor
L	Length of metal sheet (mm)
Nu	Nusselt number
Pr	Prandtl number
Re	Reynolds number
UA	Heat transfer performance (W/K)
h	Heat transfer coefficient (W/m2·K)
k	Thermal conductivity (W/m·K)
N	Stacked number of metal sheet
ΔP	Pressure drop (kPa)
Q	Heat transfer rate (W)
ΔTLMTD	Log mean temperature difference (K)
W	Width of metal sheet (mm)

Greek Symbols

ρ Fluid density (kg/m3)

μ Dynamic viscosity (N·s/m2)

Π Uncertainty

Subscripts

c Cold

i Inlet

o Outlet

h Hot

m Mean

REFERENCES

1. Kim, Y.H. Heat Transfer and Fluid Flow Characteristics on Printed Circuit Heat Exchangers with Microchannel Configurations. Ph.D. Thesis, Korea University, Seoul, Korea, 2009.
2. Johnston, A.M.; Levy, W.; Rumbold, S.O. Application of PCHE technology within heterogeneous catalytic reactors. Proceedings of the American Institute of Chemical Engineers (AIChE) Annual Meeting, Reno, NV, USA, 4–9 November 2001.

3. Gromoll, B. Micro cooling systems for high density packaging. Revue Générale de Thermique 1998, 37, 781–789.

4. Kwon, O.K.; Choi, M.J.; Choi, Y.J. Heat transfer and pressure drop characteristics in zigzag channel angles of printed circuit heat exchangers. Korean J. Air-Cond. Refrig. Eng 2009, 21, 475–482.

5. Nikitin, K.; Kato, Y.; Ngo, N. Printed circuit heat exchanger thermal-hydraulic performance in supercritical CO2 experimental loop. Int. J. Refrig 2006, 29, 807–814.

6. Peng, X.F.; Peterson, G.P. Convective heat transfer and flow friction for water flow in microchannel structures. Int. J. Heat Mass Transf 1995, 39, 2599–2608.

7. Lee, J.; Mudawar, I. Assessment of the effectiveness of nanofluids for single-phase and two-phase heat transfer in micro-channels. Int. J. Heat Mass Transf 2007, 50, 452–463.

8. Qu, W.; Mudawar, I. Experimental and numerical study of pressure drop and heat transfer in a single-phase micro-channel heat sink. Int. J. Heat Mass Trans 2002, 45, 2549–2565.

9. Shen, S.; Xu, J.L.; Zhou, J.J.; Chen, Y. Flow and heat transfer in microchannels with rough wall surface. Energy Convers. Manag 2006, 47, 1311–1325.

10. Rachkovskij, D.A.; Kussul, E.M.; Talayev, S.A. Heat exchange in short microtubes and micro heat exchangers with low hydraulic losses. Micro. Technol 1998, 4, 151–158.

11. Kang, S.W.; Tsen, S.C. Analysis of effectiveness and pressure drop in micro cross-flow heat exchanger. Appl. Therm. Eng 2007, 5–6, 877–885.

12. Ngo, L.; Katoa, Y.; Nikitina, K.; Tsuzukia, N. New printed circuit heat exchanger with S-shaped fins for hot water supplier. Exp. Therm. Fluid Sci 2006, 30, 811–819.

13. Tsuzuki, N.; Kato, Y.; Ishizuka, T. High performance printed circuit heat exchanger. Appl. Therm. Eng 2007, 30, 1702–1707.

14. Ma, T.; Zeng, M.; Luo, T. Numerical study on thermo-hydraulic performance of an offset-bubble primary surface channels. Appl. Therm. Eng 2013, 61, 44–52.

15. Ma, T.; Xin, F.; Li, L. Effect of fin-endwall fillet on thermal hydraulic performance of airfoil printed circuit heat exchanger. Appl. Therm. Eng 2015.

16. Ma, T.; Li, L.; Xu, X.; Chen, Y.; Wang, Q. Study on local thermal-hydraulic performance and optimization of zigzag-type printed circuit heat exchanger at high temperature. Energy Convers. Manag 2015.

17. Baek, S.W.; Lee, C.K.; Jeong, S.K. Effect of flow maldistribution and axial conduction on compact microchannel heat exchanger. Cryogenics 2014, 60, 49–61.

18. Bartel, N.; Chen, M.; Utgikar, V.P. Comparative analysis of compact heat exchangers for application as the intermediate heat exchanger for advanced nuclear reactors. Ann. Nucl. Energy 2015, 81, 143–149.

19. Figley, J.; Sun, X.; Mylavarapu, S.K. Numerical study on thermal hydraulic performance of a Printed Circuit Heat Exchanger. Prog. Nucl. Energy 2013, 68, 89–96.

20. Kim, T.H.; Kwon, J.G.; Yoon, S.H. Numerical analysis of air-foil shaped fin performance in printed circuit heat exchanger in a supercritical carbon dioxide power cycle. Nucl. Eng. Des 2015, 288, 110–118.

21. Koo, G.W.; Lee, S.M.; Kim, K.Y. Shape optimization of inlet part of a printed circuit heat exchanger using surrogate modeling. Appl. Therm. Eng 2014, 72, 90–96.

22. Mylavarapu, S.K.; Sun, X.; Glosup, R.E. Thermal hydraulic performance testing of printed circuit heat exchangers in a high-temperature helium test facility. Appl. Therm. Eng 2014, 65, 605–614.

23. Xu, X.; Ma, T.; Li, L. Optimization of fin arrangement and channel configuration in an airfoil fin PCHE for supercritical CO2 cycle. Appl. Therm. Eng 2014, 70, 867–875.

24. Yoon, S.H.; NO, H.C.; Kang, G.B. Assessment of straight, zigzag, S-shape, and airfoil PCHEs for intermediate heat exchangers of HTGRs and SFRs. Nucl. Eng. Des 2014, 270, 334–343.

25. Yoon, S.J.; Sabharwall, P.; Kim, E.S. Numerical study on crossflow printed circuit heat exchanger for advanced small modular reactors. Int. J. Heat Mass Transf 2014, 70, 250–263.

26. Jeong, C.H.; Kim, H.R.; Ha, M.Y. Numerical investigation of thermal enhancement of plate fin type heat exchanger with creases and holes in construction machinery. Appl. Therm. Eng 2014, 62, 529–544.

27. Kim, I.H.; Sun, X. CFD study and PCHE design for secondary heat exchangers with FLiNaK-Helium for SmAHTR. Int. J. Nucl. Eng. Des 2014, 270, 325–333.

28. Kim, I.H.; NO, H.C. Thermal hydraulic performance analysis of a printed circuit heat exchanger using a helium-water test loop and numerical simulations. Appl. Therm. Eng 2011, 31, 4064–4073.

29. Kim, I.H.; NO, H.C. Thermal-hydraulic physical models for a Printed Circuit Heat Exchanger covering He, He-CO2 mixture, and water fluids using experimental data and CFD. Exp. Therm. Fluid Sci 2013, 48, 213–221.

30. Cowell, T.A. A general method for the comparison compact heat transfer surfaces. J. Heat Transf. ASME 1990, 112, 288–294.

31. Shah, R.K. Assessment of modified Wilson plot techniques for obtaining heat exchanger design data. Heat Transf 1990, 5, 51–56.

32. Kays, W.M.; London, A.L. Compact Heat Exchangers, 2nd ed; McGraw-Hill: New York, NY, USA, 1964.

33. ANSI/ASME PTC 19.1, Measuring Uncertainty; The American Society of Mechanical Engineers: New York, NY, USA, 1998.

34. Taylor, B.N.; Kuyatt, C.E. Guidelines for Evaluating and Expressing the Uncertain of NIST Measurement Results; NIST Technical Note 1297; NIST: Gaithersburg, MD, USA, 1994.

35. Wilson, E.E. A basis for rational design of heat transfer apparatus. J. Heat Transf. ASME 1915, 37, 47–82.

36. Manglik, R.M.; Bergles, A.E. Heat Transfer Enhancement of Intube Flows in Process Heat Exchangers by Means of Twisted-tape Inserts; Report No. HTL-18; Heat Transfer Laboratory, Rensselaer Polytechnic Institute: Troy, NY, USA, 1991.

37. Hesselgreaves, J.E. Compact Heat Exchangers; Pergamon: Edinburgh, UK, 2001.

2

CHAPTER

CONDENSATION HEAT TRANSFER IN HORIZONTAL NON-CIRCULAR MICROCHANNELS

Hicham El Mghari[1,2], Mohamed Asbik[1], Hasna Louahlia-Gualous[2]

[1]*LP2MS, URAC08, Université Moulay Ismaïl, Faculté des Sciences, Meknès, Maroc*

[2]*LUSAC, Université de Caen Basse Normandie, Saint Lô, France*

ABSTRACT

This investigation contributes to a better understanding of condensation heat transfer in horizontal non-circular microchannels. For this purpose, the conservation equations of mass, momentum and energy have been numerically solved in both phases (liquid and vapor), and all the more, so the film thickness analytical expression has been established. Numerical results relative to variations of the meniscus curvature radius, the condensate film thickness, the condensation length and heat transfer coefficients, are analyzed in terms of the influencing physical and geometrical quantities. The effect of the microchannel shapes on the average Nusselt number is highlighted by studying condensation of steam insquare, rectangular and equilateral triangular microchannels with the same hydraulic diameter of 250 μm.

Keywords: Condensation; Microchannel; Numerical simulation; Capillary regime; Heat transfer

1. INTRODUCTION

Understanding the heat transfer behavior of condensation flow in microchannels is important for a broad variety of engineering applications. Although there have been a number of investigations on boiling flow in microchannels, there are relatively little experimental data and theoretical analyses relative to condensation processes available in the literature, especially, for condensation inside a noncircular microchannels. In the chapter 6 of the reference [1], an overview of minichannels and microchannels condensation has been exposed.

Most of the physical and mathematical models that focused on annular condensation heat transfer in circular channel were developed in the previous works. Begg et al. [2] studied annular film condensation in a small circular tube to predict the shape of the liquid-vapor interface along a miniature tube leading to the complete condensation phenomena in small diameter tubes. LouahliaGualous and Asbik [3] conducted a numerical model predicting heat transfer for condensation of pure refrigerant and binary mixture in a mini-tube. Miscevic et al. [4] developed a stationary condensation capillary flow model based on the separate flow approach by taking into account the coupling between a cylindrical interface and a hemispherical interface. Recently, Ribeiro et al. [5] experimentally investigated the thermal-hydraulic performance of microchannel condensers using three different copper metal foams structures with distinct pore densities and porosities (0.893 and 0.947) as enhanced surfaces on the air-side. Their results are compared with the conventional condenser surface. El Achkar et al. [6] investigated the experimental heat transfer in the isolated bubbles zone of a transparent circular cross-section micro condenser. The evolution of vapor quality was experimentally determined by using the image processing. The energy balance was then used to calculate the temperature of the liquid in the isolated bubbles zone, showing that liquid and vapor were not in thermal equilibrium. The sensible heat transfers and latent heat transfers were then compared. However, a fundamental understanding of local mechanisms of heat and mass transfer cannot be accomplished using multichannels because the mass flow rates for each microchannel and condensate flow regime are unknown. Also, condensation heat transfer and pressure drop depend on the corresponding structure of the two phase flow (mist flow, annular flow, bubbly flow, or slug/plug flow) as shown by Odaymet et al. [7]. Three main condensation flows were identified in a small circular tube: the annular flow, the intermittent or elongated bubbles flow, and the spherical bubbles flow by Louahlia-Gualous and Mecheri [8]. Annular flow is especially found to be one of the dominant condensation flows in microchannelsas shown by Odaymet and LouahliaGualous [9] and Quan et al. [10].

On the other hand, various theoretical models have been proposed to predict the local heat transfer related to the condensation annular flow in non-circular channel where the surface tension plays a predominant effect on the condensate flow, more specifically, in the channel corners. Indeed, Zhao and Liao [11] analyzed annular film condensation heat transfer inside vertical mini triangular channel using three zones: the thin liquid film flow on the sidewall, the condensate flow in the corners, and the vapor core flow in the center. Wu and Cheng [12] carried out a simultaneous visualization and measurement experiment to perform condensation flow patterns of steam flowing through an array of trapezoidal silicon microchannels with a hydraulic diameter of 82.8 μm. Wang et al. [13] and Wang & Rose [14] proposed a theoretical model for condensation annular flow in a horizontal square and equilateral triangular channel with hydraulic diameter ranging from 0.5 to 5 mm by taking into account the effects of gravity, surface tension, and interfacial shear stress. They obtained the local heat transfer coefficient for refrigerants R134a, R22, R152a, CO2, propane, ammonia, and R410a by assuming that the channel wall temperature is uniform. Furthermore, they proposed one correlation for condensation heat transfer in the square and triangular microchannels in which surface tension and viscosity are the predominant parameters controlling condensate film thickness. Additionally, in references ([15-18]) for which the contents are not detailed here, readers could find more information about the use of various non-circular microchannel shapes.

The main purpose of the present work is to determine the heat transfer coefficient during the steam condensation inside horizontal various non-circular microchannels (rectangle, square, or equilateral triangle). Indeed, the classical mathematical model of the annular condensation flow is retained in both phases (liquid and vapor). An appropriate numerical method is used to solve the differential equations system obtained from different conservation equations (mass, momentum and energy). To compute the heat transfer coefficient, a new and simple geometrical method is used to express the condensate film thickness.

2. PHYSICAL MODEL

The physical model investigated in this paper is illustrated on the Figure 1. It concerns, a horizontal non circular microchannel with a hydraulic diameter D, which is cooled with a uniform wall heat flux density q. At the microchannel inlet, the vapor (steam) mass flux, the pressure and temperature of the steam is given at the saturated state. The vapor condenses inside the microchan nel having wall temperature lower than vapor saturation temperature. The film thickness on the heat exchange surface varies along the axial direction with the vapor quality.

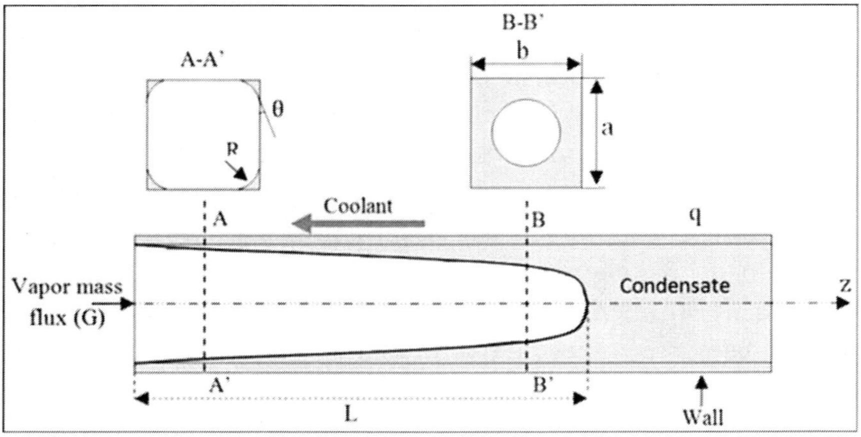

Figure 1. Schematic representation of the annular condensation in a rectangular microchannel.

When total condensation occurs, the end of the condensation zone has a hemispherical meniscus. Its length L is one of the unknown parameters in the physical model. The condensate film flows along the axial direction under effects of the pressure, surface tension, and shear stress.

3. MATHEMATICAL FORMULATION

3.1. Hypotheses

The mathematical formulation of the problem is based on the following principal assumptions:

- For liquid and vapor phases, thermophysical properties are assumed to be constant.

- In the microchannel, the flow is supposed to be steady-state, laminar, one-dimensional and axis-symmetrical.
- The free surface of the condensate film is smooth.
- Gravity forces are negligible compared to the effects of surface tension.
- The heat transfer from the cooling fluid to the condensate flow is assumed to be one-dimensional.
- In the condensate, the temperature profile is supposed linear.
- The saturation temperature of the vapor is assumed to remain constant along the microchannel.

3.2. Conservation Equations

The modeling approach developed here describes the liquid and vapor phases separately. The governing equations are used in the Cartesian coordinates as shown in the Figure 1.

3.2.1. Mass Conservation

The average parameters over a cross-section are used in liquid and vapor phases of the condensation flow respecting continuity conditions at the liquid-vapor interface. The equation of the mass conservation can be written for each local cross section as follows:

$$\frac{1}{A}\frac{\partial}{\partial z}\left(\rho_l U_l (1-\alpha) A\right) = -\Gamma \tag{1}$$

$$\frac{1}{A}\frac{\partial}{\partial z}\left(\rho_v U_v \alpha A\right) = \Gamma \tag{2}$$

where Γ represents the volumic rate of phase change. By convention, its sign is negative for condensation and positive for evaporation. So, we can write:

$$\Gamma = -\mathrm{d}\dot{m}_L / \mathrm{d}z = \mathrm{d}\dot{m}_v / \mathrm{d}z \tag{3}$$

A_ξ, A_ξ and U_ξ are respectively the microchannel cross section, the density and the axial velocity. The subscript ξ refers to the considered phase (ξ = v or L).

3.2.2. Momentum Conservation

For a fixed position z along the microchannel, the condensate film thickness in noncircular microchannels is much thicker into the microchannel corners than elsewhere, especially; at the internal considered circumference because of the surface tension effect. This is the reason that the axial flow in the film region between the corners is neglected.

From the forces balance illustrated on the Figure 2, the momentum conservation equations in the control volume of the length dz, are given by the equations (4) and (5). These relationships are essentially expressed in terms ofthe interfacial shear at the liquid-vapor interface, the shear wall friction at the liquid-wall contact surface, and the pressure forces on the liquid area. In both phases, all the physical properties are assumed to be constant, and the influences of the gravity and the buoyancy forces are neglected. So, in the liquid and vapor phases, the momentum conservation equations are:

- In the liquid phase:

$$-A_L \frac{\mathrm{d}P_L}{\mathrm{d}z} + \tau_{vL} \frac{\mathrm{d}S_{vL}}{\mathrm{d}z} - \tau_w \frac{\mathrm{d}S_{Lw}}{\mathrm{d}z} - U_L \frac{\mathrm{d}\dot{m}_L}{\mathrm{d}z} = \dot{m}_L \frac{\mathrm{d}U_L}{\mathrm{d}z} \quad (4)$$

- In the vapor phase:

$$-A_v \frac{\mathrm{d}P_v}{\mathrm{d}z} + \tau_{vL} \frac{\mathrm{d}S_{vL}}{\mathrm{d}z} - U_v \frac{\mathrm{d}\dot{m}_v}{\mathrm{d}z} = \dot{m}_v \frac{\mathrm{d}U_v}{\mathrm{d}z} \quad (5)$$

A_ξ, A_ξ and U_ξ are respectively the microchannel cross-section, the pressure and the axial velocities in the phase $\xi(v \text{ or } L)$. Furthermore, we indicate that $\mathrm{d}S_{vL}$ is the liquid–vapor interface surface along $\mathrm{d}z$, $\mathrm{d}S_{Lw}$ is the wet heat exchange surface along $\mathrm{d}z$, and τ_{vL} is the shear stress at liquid–vapor interface, τ_w is the shear stress at the microchannel heat exchange surface.

3.2.3. Energy Conservation

The local energy equation in the liquid phase can written as:

$$q\mathcal{P}z = \dot{m}_L(z) h_{fg} \quad (6)$$

The total energy equation defined in the length of the total condensation zone as:

$$q\mathcal{P}L = \dot{m}_{v,in} h_{fg} \quad (7)$$

where q is the heat flux density, $\dot{m}_L(z)$ is the local liquid mass flow rate $\mathcal{P}$ is the microchannel perimeter, and z is the microchannel abscissa.

3.2.4. Curvature Radius Expression

To express the curvature radius derivative, we need to use the Laplace–Young equation:

$$\frac{dP_L}{dz} = \frac{dP_v}{dz} + \frac{\sigma}{R^2}\frac{dR}{dz} \tag{8}$$

Combining equations (1)-(8), we get:

$$\frac{dR}{dz} = \frac{0.5 f_{vL}\rho_v \mathcal{P}_{vL} U_v^2 \left[1 + \frac{A_L}{A_v}\right] - 0.5 f_L \rho_L \mathcal{P}_{Lw} U_L^2 - \frac{2q\mathcal{P}A_L}{h_{fg}}\left[\frac{U_v}{A_v} + \frac{U_L}{A_L}\right]}{\sigma \frac{A_L}{R^2} + 2\frac{\rho_v U_v^2 A_L^2}{A_v R} - \frac{2\rho_L A_L}{R} U_L^2} \tag{9}$$

3.2.5. Dimensionless Equations

To establish the dimensionless equations, the following variables are used:

$$R^* = \frac{R}{D_h}, z^* = \frac{z}{D_h}, L^* = \frac{L}{D_h}, \bar{\rho} = \frac{\rho_v}{\rho_L}, \mathcal{P}^* = \frac{\mathcal{P}}{D_h}, P^* = \frac{PD_h}{2\sigma}, U_V^* = \frac{A\rho_v U_v}{\dot{m}_{v,in}}, U_L^* = \frac{A\rho_L U_L}{\dot{m}_{v,in}} \tag{10}$$

Then the dimensionless form of the derivative curvature radius, the velocity gradients and the pressure gradients in both phases, are given by:

$$\frac{dR^*}{dz^*} = \frac{0.5 f_{vL} \mathcal{P}_{vL}^* U_v^{*2}\left[1 + \frac{A_L^*}{A_v^*}\right] - 0.5 f_L \rho^* \mathcal{P}_{Lw}^* U_L^{*2} - 2Bo\mathcal{P}^* A_L^* \left[\rho^* \frac{U_L^*}{A_L^*} + \frac{U_v^*}{A_v^*}\right]}{\frac{A_L^* \rho^*}{Re\, Ca} + \frac{2A_L^{*2} U_v^{*2}}{A_v^* R^*} - \frac{2\rho^* A_L^* U_L^{*2}}{R^*}} \tag{11}$$

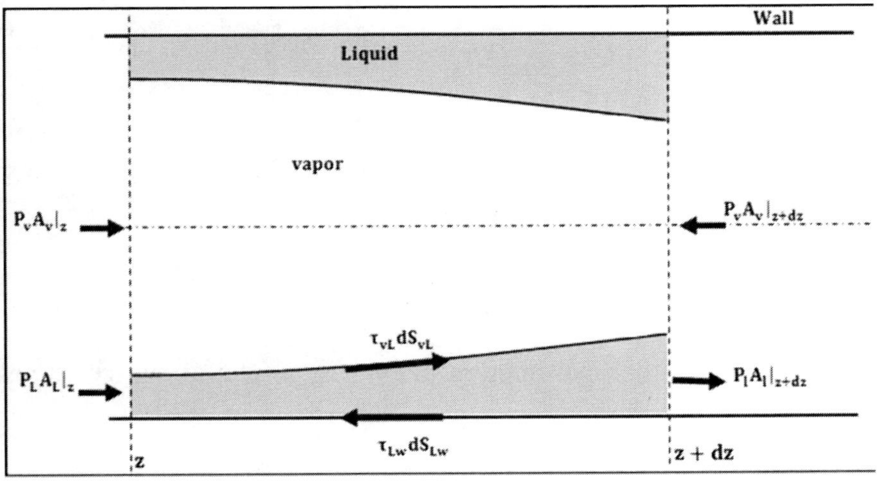

Figure 2. Forces balance for two-phase flow in a horizontal microchannel.

$$\frac{\mathrm{d}U_L^*}{\mathrm{d}z^*}=\frac{Bo\mathcal{P}^*}{A_L^*}-2\frac{U_L^*}{R^*}\frac{\mathrm{d}R^*}{\mathrm{d}z^*} \tag{12}$$

$$\frac{\mathrm{d}U_v^*}{\mathrm{d}z^*}=-\frac{Bo\mathcal{P}^*}{A_v^*}-2\frac{A_L^*}{A_v^*}\frac{U_v^*}{R^*}\frac{\mathrm{d}R^*}{\mathrm{d}z^*} \tag{13}$$

$$\frac{\mathrm{d}P_L^*}{\mathrm{d}z^*}=\frac{1}{4}\frac{f_{vL}Ca\,Re\mathcal{P}_{vL}^*U_v^{*2}}{\bar{\rho}A_L^*}-\frac{1}{4}\frac{f_LCa\,Re\mathcal{P}_{Lw}^*U_L^{*2}}{A_L^*}-\frac{Bo\;Ca\;Re\mathcal{P}^*U_L^*}{A_L^*}+\frac{Ca\;ReU_L^{*2}}{R^*}\frac{\mathrm{d}R^*}{\mathrm{d}z^*} \tag{14}$$

$$\frac{\mathrm{d}P_v^*}{\mathrm{d}z^*}=-Ca\;ReU_v^*\frac{\mathrm{d}U_v^*}{\mathrm{d}z^*}+\frac{Ca\;Re}{A_v^*}\frac{\mathrm{d}A_v^*}{\mathrm{d}z^*}+\frac{Ca\;ReU_v^{*2}}{R^*}\frac{A_L^*}{A_v^*}\frac{\mathrm{d}R^*}{\mathrm{d}z^*}-\frac{1}{4}\frac{f_{vL}Ca\;Re\mathcal{P}_{vL}^*U_v^{*2}}{\bar{\rho}A_v^*} \tag{15}$$

$$Bo = q/Gh_{fg}\ ,\ \ Ca = \frac{\mu_l G}{\sigma \rho_L},$$

$$Re = \frac{\dot{m}_{v,in} D_h}{A\mu_L},\ \ \dot{m}_{v,in} = GA$$

are respectively the boiling number, the capillary number, the vapor Reynolds number and the vapor mass flux.

Definitions of other parameters appearing in the above relations (11-15) are such that:

1) The liquid friction factor for laminar is given by:

$$f_L = CRe_L^{-1} \tag{16}$$

C is the Poiseuille number given in [19].

For turbulent flow, the liquid friction coefficient is determined from the Blasius equation [1]:

$$f_L = 0.0791\,\mathrm{Re}_L^{-0.25} \tag{17}$$

In this expression, the liquid Reynolds number Re_L is calculated assuming the liquid single phase:

$$Re_L = \frac{\rho_L U_L D_L}{\mu_L},\ \ D_L = \frac{4A_L}{P_{Lw}} \tag{18}$$

2) The interfacial frictional coefficient taking into account the effect of the condensation process on the interfacial shear stress is defined by [20]:

$$f_{vL} = f_{v,0}\frac{\varnothing \mathrm{e}^{\varnothing}}{\mathrm{e}^{\varnothing} - 1} \tag{19a}$$

where $f_{v,0}$ is the friction factor for single phase vapor flow defined for laminar flow as :

$$f_{v,0} = CRe_v^{-1} \tag{19b}$$

For turbulent flow, $f_{v,0}$ is determined from:

$$f_{v,0} = 0.0791Re_v^{-0.25} \tag{20a}$$

$$Re_v = \frac{\rho_v U_v D_v}{\mu_v},\ D_v = \frac{4A_v}{\mathcal{P}_{vL}} \quad , \tag{20b}$$

The factor $\varnothing$ is defined as the ratio of the local condensation mass flow rate to the vapor mass flow rate rebounding from the liquid-vapor interface [2], which is approximated by:

$$\varnothing = \frac{dq}{\mathcal{P}dz}\frac{2}{h_{fg}U_v\rho_v f_{v,0}} \tag{21}$$

3.2.6. Dimensionless Boundary Conditions

Equations (11) to (15) are solved using the following dimensionless boundary conditions:

1) At the microchannel inlet ($z^* = 0$):

- the flow mass flux G is imposed;
- the temperature and pressure at the saturated state are given;
- the non-dimensional vapor velocity is

$$U_v^*\left(z^* = 0\right) = 1;$$

- the non-dimensional liquid velocity is

$$U_L^*\left(z^*=0\right)=0$$;

- the non-dimensional curvature radius is

$$R^*\left(z^*=0\right)=\frac{\sigma}{P_{v0}D_h}$$

2) At the position z* = L* corresponding to the end of the condensation zone:

- the non-dimensional outlet liquid pressure is:

$$P_l\left(z^*=L^*\right)=\frac{P_0 D}{2\sigma}$$;

- for the rectangular cross section, the dimensionless curvature radius is [22]:

$$R^*\left(z^*=L^*\right)=\varepsilon\sin(\beta)/\left(2\cos(\theta+\beta)D_h\right)$$

where $\varepsilon=\min(a,b)$;

- for the equilateral triangle cross section, the dimensionless curvature radius a is [18] :

$$R^*\left(z^*=L^*\right)=\mathcal{P}\sin(\beta)/\left(6\cos(\theta+\beta)D_h\right)$$

4. SOLUTION PROCEDURE

The above boundary conditions combined with nondimensional equations (11)-(15) constitute the mathematical model of two-phase flow in capillary regime with a vapor-liquid phase change. Solution of this mathematical model is not trivial since one of the limit positions remains to be found, the value of L being one of the unknowns of the problem. So, the iteration process is used to solve the mathematical model. To start the calculations, the saturated vapor at the inlet of the tube is assumed having the known mass flux, temperature, pressure, and vapor quality. For the next axial step $\left(z^* + \Delta z^*\right)$, the following steps are executed:

1) To start calculation, an arbitrary total condensation length is assumed;
2) Calculation of the curvature radius from equation (11);
3) Calculation of the local liquid velocity from equation (12);
4) Calculation of the local vapor velocity from equation (13);
5) Calculation of the liquid pressure from equation (14);
6) Calculation of the local vapor pressure from equation (15);
7) Calculation of the total condensation length from equation (26).

Steps 2) to 7) are repeated until the value of the condensation length obtained at the iteration number "it" is approximately equal to the one determined at iteration number "it-1".

Knowing the total condensation length, calculations are then made for the next values of z locations. For each z location, steps 2) to 7) are made and calculation is stopped when z value is approximately equal to the total condensation length L.

5. COMPUTATION OF THE HEAT TRANSFER COEFFICIENT AND THE CONDENSATION LENGTH

5.1. Calculation of the Heat Transfer Coefficient

Assuming that the temperature profile is linear in the condensate film, the local heat transfer is expressed by:

$$h(z) = \frac{\lambda_L}{\delta(z)} \tag{22a}$$

and hence the average heat transfer coefficient in the microchannel condensation length L is performed by the relationship:

$$h_m = \frac{1}{L}\int_0^L h(z)\,\mathrm{d}z \tag{22b}$$

Two last quantities are used to determine the average and the local Nusselt numbers respectively:

$$Nu_{av} = \frac{h_m D}{\lambda_L} \tag{23a}$$

$$Nu_{loc} = \frac{h(z) D}{\lambda_L} \tag{23b}$$

$\delta(z)$ is the average condensate film thickness for each z location along the microchannel. Using the geometrical considerations from the Figure 3, the following analytical expression is established:

$$\delta(z) = \frac{1}{\frac{\pi}{2}}\int_0^{\frac{\pi}{2}} R\left[\frac{\cos(\theta)}{\cos(\Omega)} - 1\right]\mathrm{d}\Omega \tag{24}$$

5.2. Calculation of the Condensate Length

The parameter dR/dz given by the equation (9) is expressed in terms of the limit curvature radius R_L.

At z = L location, dR/dz is to be infinite which causes a calculation problem. To avoid this complication, we set $z = L - \varepsilon$ where ε is an infinitesimal parameter. The limit values of the curvature radius and its first derivative are defined as [22]:

$$R_L = R\big|_{z=L-\varepsilon} = \frac{\varepsilon \sin\beta}{2\cos(\theta+\beta)D_h} \tag{25}$$

$$\left.\frac{\mathrm{d}R}{\mathrm{d}z}\right|_{z=L-\varepsilon} = \frac{\varepsilon\cos\theta - R_L}{R_L\cos\theta} \tag{26}$$

To estimate the numerical value of the condensation length a dichotomy method was performed using equation (26), between the inlet of the microchannels which corresponds to $(z = 0 \text{ and } R = 0)$ and the end of the condensation zone where $(z = L - \varepsilon \text{ and } R = R_L)$.

6. RESULTS AND DISCUSSIONS

6.1. Numerical Results Validation

The validation of the results obtained in the present work for condensation of water in a square microchannel, are compared to the predictions of various correlations available in the literature. These correlations are proposed for condensation heat transfer in microchannels and macrochannels. Among these predictive correlations, those of condensation in microchannels are defined by Wang et al. [26], Koyama et al. [23], and Wang & Rose [14]. Figure 4

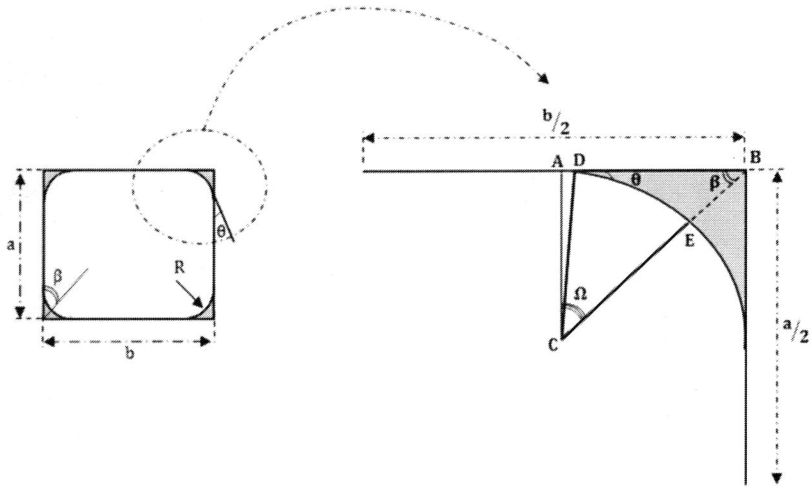

Figure 3. Distribution of the condensate film in the microchannel corners.

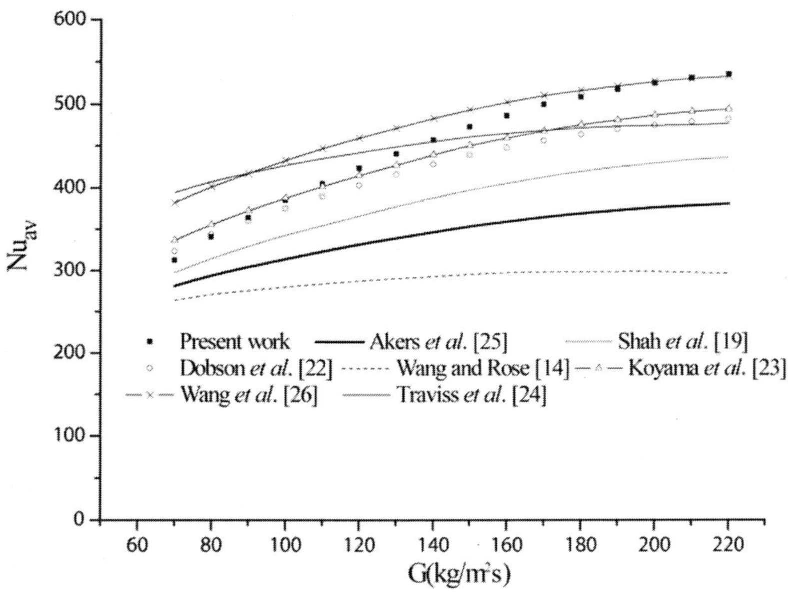

Figure 4. Comparison of the present work with predictions of literature correlations.

shows the comparisons between the average Nusselt number obtained from the present numerical model and those predicted by seven correlations: Dobson et al. [22], Wang et al. [26], Koyama et al. [23], Wang & Rose [14], Traviss et al. [24], Shah [19], and Ackers et al. [25]. Results presented in the Figure 4 are those obtained for condensation of steam in a horizontal square microchannel with hydraulic diameter of 110 μm, for various inlet-vapor mass fluxes ranging from 70 to 220 kg/m^{2}s. It is found that the best predictions of the present average Nusselt number are obtained by the correlations of Dobson et al. [22] and Koyama et al. [23] for which the mean relative deviation is about 7%. Both of these correlations are developed in the case of an annular condensation heat transfer of several pure refrigerants which is carried out in the channels with small hydraulic diameter. The correlations of Traviss et al. [24] and Ackers et al. [25] devoted to the convective condensation heat transfer in the macro-scale channels have been also evaluated to the micro-scale. It can be seen that the correlation of Traviss et al. [24] over predicts highly the present results for low mass fluxes whereas the Ackers's correlation [25] under predicts them highly for all the range of the tested mass fluxes. Adding up, the correlation of Shah [19] is proposed for condensation heat transfer in channels with large hydraulic diameter from 7 to 40 mm and Wang et al. [26] correlation is defined for condensation heat transfer inside a horizontal small rectangular channel with hydraulic diameter of 1.46mm. Average Nusselt numbers obtained by both of these correlations give the same trend and the reasonable results compared to those predicted numerically. The Shah's correlation under predicts numerical results with a maximum average deviation of about 16%. As for the correlation of Wang et al. [26], it over predicts the present results with a maximum deviation of 22% obtained at low mass fluxes. Moreover, correlation of Wang & Rose [14] based on the Nusselt theory including the effects of interfacial shear stress and surface tension on condensation heat transfer is also evaluated. Predictions of Nusselt number from the correlation of

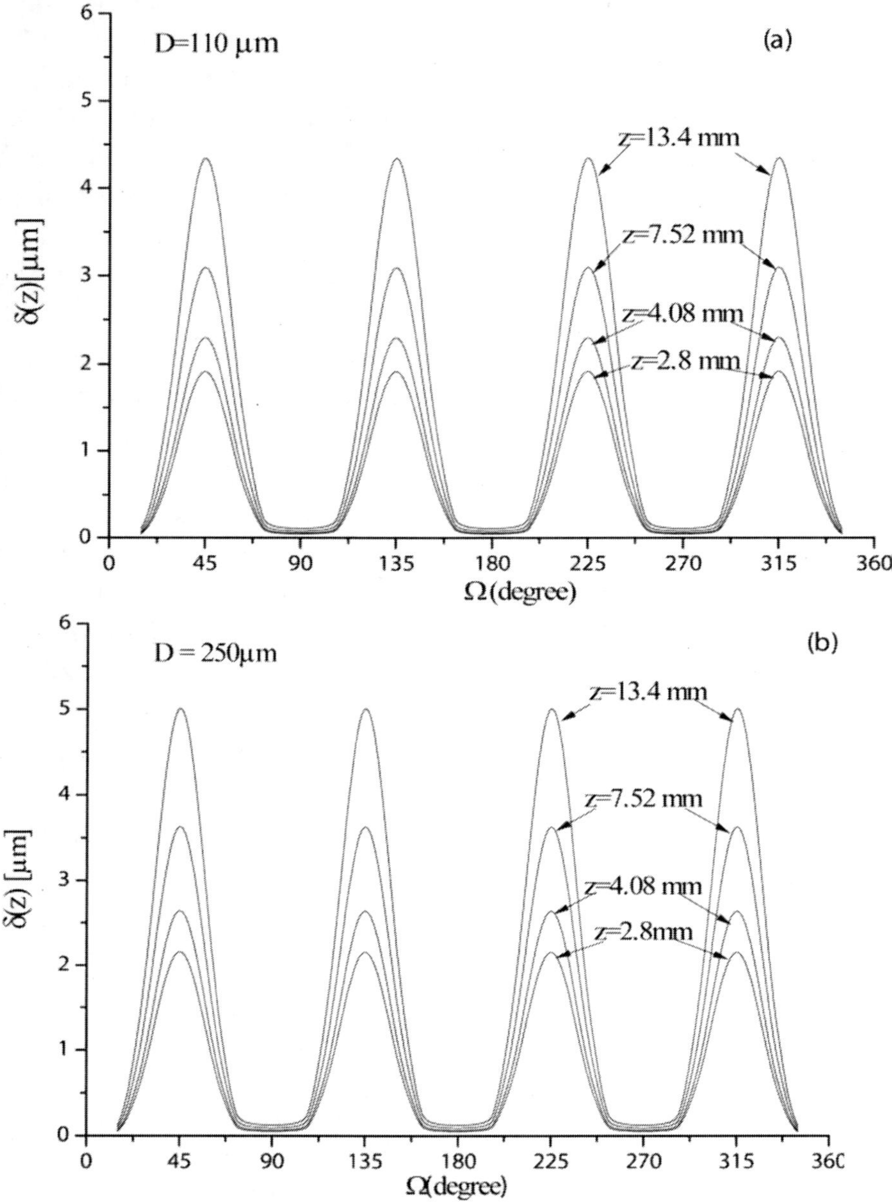

Figure 5. Evolution of the condensate film thickness around the square microchannel circumference.

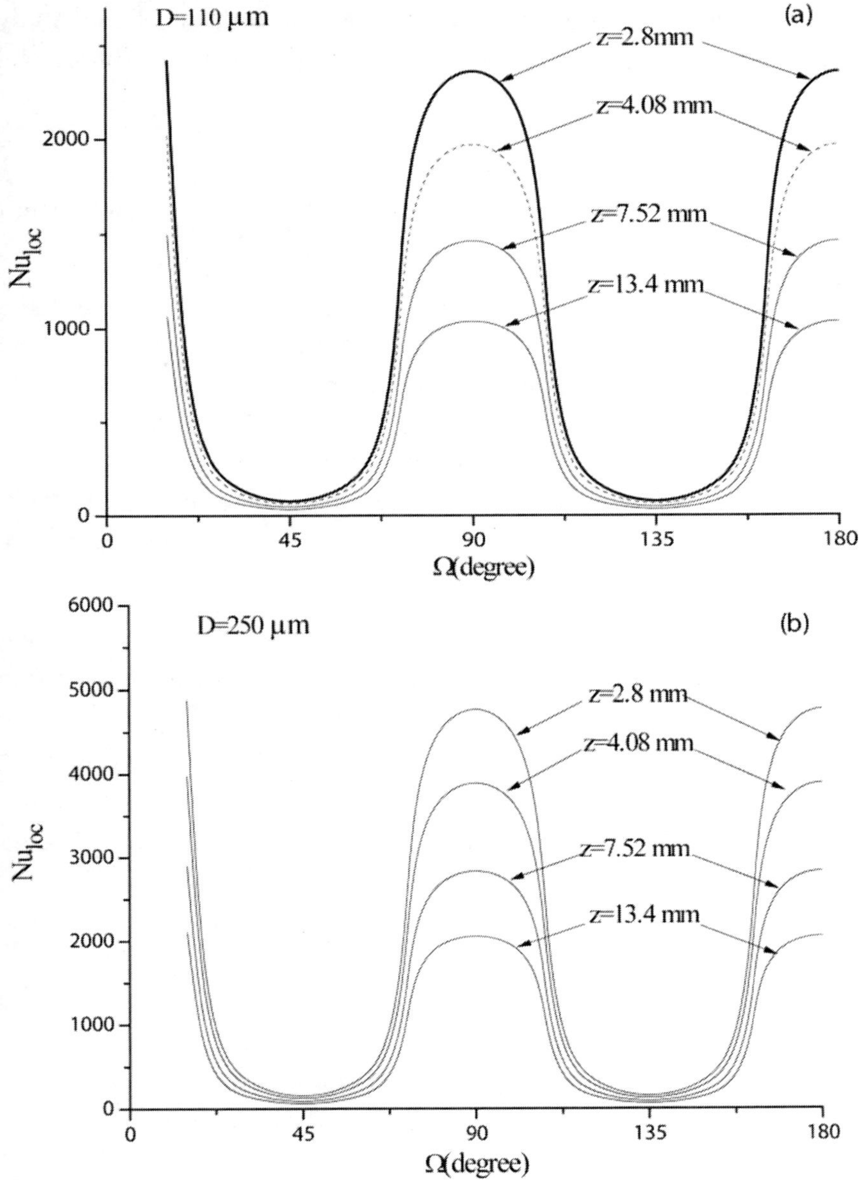

Figure 6. Evolution of the local Nusselt number around the square microchannel circumference.

Wang & Rose [14] are not in accordance with those of this study since they gave the highest deviation of 37% form the numerical results.

6.2. Peripheral Condensate Film Thickness and Local Heat Transfer

Numerical results are given in the Figures 5(a) and (b) and the Figures 6(a) and (b) for steam condensation in square section microchannel with hydraulic diameter of 110 and 250 μm. Computations were conducted for vapor mass flux $G = 90$ kg/m^{2}s, contact angle $q = 15°$, heat flux density $q = 100$ kW/m^2. For the conditions used in the present computation, the boiling number, the capillary number, the inlet steam temperature and pressure are maintained constant.

To better understand the behavior of the peripheral local heat transfer coefficient, Figures 5(a) and (b) show predicted square channel condensate film thickness (24) plotted as a function of the peripheral coordinate W at four different axial locations. It can be seen from these figures that for the same z location on the microchannel, the condensate film thickness is higher for 250 μm (Figure 5(b)) than for 110 μm (Figure 5(a)). According to these representations, it is observed that for both microchannels, the symmetrical distributions of the condensate film thickness are obtained around the channel perimeter because the gravity has no effect on the condensate flow in microchannel. The condensate film thickness is very thick in the microchannel corners ($W = 45°$, 135°, 225° and 315°) under the effect of the surface tension whereas it becomes very thin at $W = 0°$, 90°, 180° and 360 where the heat transfer is the highest. These trends have been reversed for the peripheral local heat transfer coefficient (see Figures 6(a) and (b)) even if the symmetrical distributions are conserved.

6.3. Influence of the Cross-Section Shape on the Heat Transfer Coefficient

To indicate the influence of the channel cross-section shape, we start by the presentation of the results concerning curvature radius. In fact, Figure 7 shows the variation of the dimensionless curvature radius along the flow direction in three cross-section shapes of microchannels. At the beginning and the end of the condensation, the curvature radius increases rapidly along the channel, while it increases slowly at the middle. We also note that for the triangular microchannel it's increasing very quickly than other geometries. The present simulation results are similar to those of the reference [21].

The annular condensation length is one of the most important parameter which influences the thermal performance of the microchannel studied here. Figure 8 gives the dimensionless annular condensation length with respect to the boiling number Bo, for three cross-section shapes. As shown in this figure, the condensation zone decreases with Bo and it is also clear that the cross-section shape of the microchannel plays a noticeable role on the condensation.

In the heat transfer exchange point of view, the influence of the microchannel shape on the average Nusselt number is highlighted by studying condensation of steam in a square, equilateral triangular, and rectangular microchannels with the same hydraulic diameter of 250 μm. The sides of the equilateral triangular and square microchannels are 433 μm and 250 μm respectively. For rectangular microchannels having the same hydraulic diameter, the aspect ratio is about 2, 3 and 4. Three different rectangular cross sections are investigated: 375 × 187.5 μm^2, 500 × 166.6 μm^2 and 625 × 156.25 μm^2.

Under the same conditions, Figure 9 compares the average Nusselt numbers with respect to the vapor Reynolds number for five microchannel cross-sections. According to this figure, it is seen that annular condensation Nusselt number is low for square microchannel cross section (aspect ratio of 1). For rectangular cross-

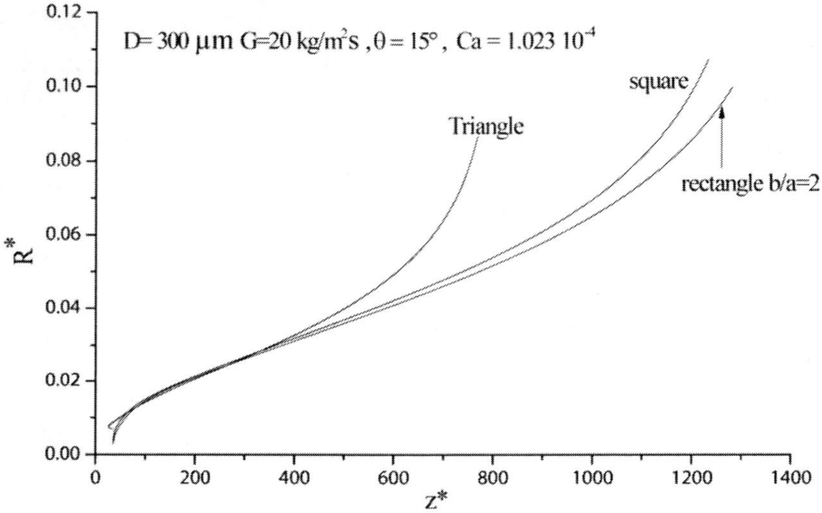

Figure 7. Dimensionless curvature radius along various microchannel shapes.

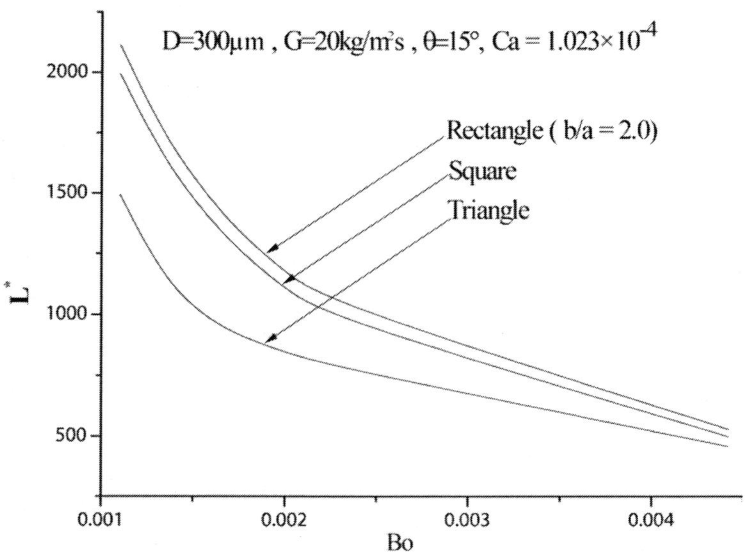

Figure 8. Dimensionless condensation length versus the boiling number.

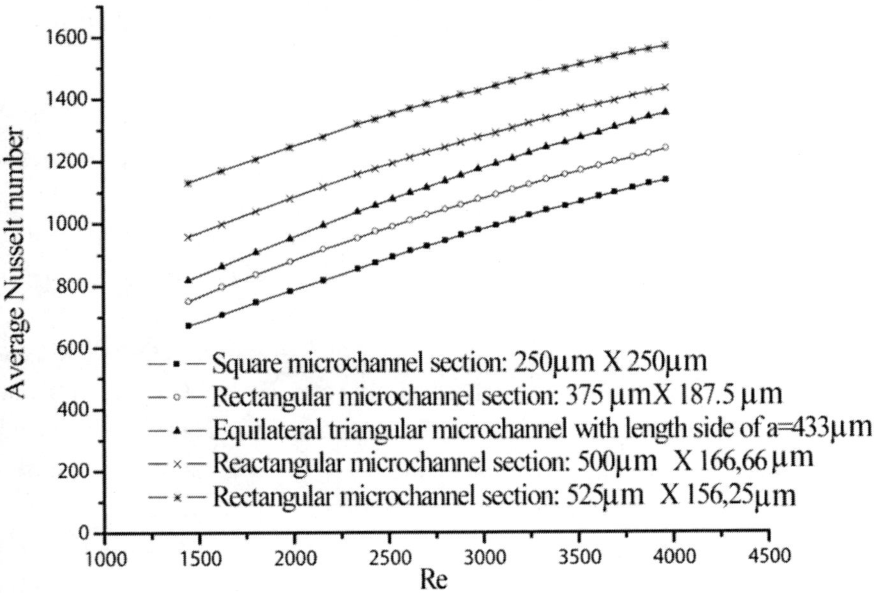

Figure 9. Comparison of the average Nusselt number for various microchannel shapes with the same hydraulic diameter (D = 250µm).

section microchannels with aspect ratio (b/a) higher than 1, condensation Nusselt number increases by increasing aspect ratio. The highest Nusselt numbers values are obtained for microchannel with high aspect ratio (equal to 4) and low confinement (a = 156.25 µm). It is interesting to note that for the same hydraulic diameter, the microchannels perimeter and cross section increase with the aspect ratio leading to a reduction of the condensate film thickness. This is due to the predominant effect of the surface tension in the microchannel that increases with microchannel perimeter and thins the condensate film along the sides of the microchannel. For the same Reynolds number, average Nusselt number for equilateral triangular microchannel is between Nusselt number for rectangular microchannel with aspect ratio of 2 and that with aspect ratio of 3. Recall that the perimeter of the triangular microchannel is about

1299 μm and those of rectangular microchannels with aspect ratios of 2 and 3 are about 1125 μm and 1333 μm respectively.

7. CONCLUSION

The numerical model characterizing local heat and mass transfer for condensation in microchannels has been developed by including the effects of wall and liquid vapor interface shear stresses, surface tension, pressure forces, contact angle, etc. Numerical results are compared for steam condensation in square, rectangular, and equilateral triangular microchannels with the same hydraulic diameter. They are compared with the correlations available in the literature and it is shown that correlations for annular condensation in macro-channels predict heat transfer with significant deviations from those obtained for microchannels. The best predictions are obtained with the correlations of Dobson et al. [25] and Koyama et al. [24].

By using the established condensate film thickness expression [24], the influence of the microchannel hydraulic diameter is highlighted. Indeed, increasing this last geometrical factor increases the peripheral condensate film thickness and hence decreases the peripheral condensation heat transfer coefficient. Additionally, we point out that at the beginning and at the end of the condensation, the curvature radius increases rapidly along the channel, while it increases slowly at the middle. We also note that for the triangular microchannel it's increasing quicker than other geometries. As for the annular condensation length, we note that it decreases with increasing of the boiling number and its numerical values diminish in the case of equilateral triangular microchannel.

Finally, to show the major influence of cross-section shape on the condensation heat transfer, comparison of the average Nusselt number is conducted for different microchannel shapes with the same hydraulic diameter. It can be concluded that condensation average heat transfer increases with aspect ratio for rectangular microchannels. The lowest average Nusselt numbers are obtained for the square microchannel because its perimeter and cross section are lower than those of triangular and rectangular microchannels.

REFERENCES

1. S. Kandlikar, S. Garimella, D. Li, S. Colin and M. King, "Heat Transfer and Fluid Flow in Minichannels and Microchannels," Elsevier, Amsterdam, 2006.
2. E. Begg, D. Khrustalev and A. Faghri, "Complete Condensation of Forced Convection Two-Phase Flow in a Miniature Tube," Journal of Heat Transfer, Vol. 121, No. 4, 1999, pp. 904-915.
3. H. Louahlia-Gualous and M. Asbik, "Numerical Modeling of Annular Film Condensation inside a Miniature Tube," Numerical Heat Transfer Part A: Applications, Vol. 52, No. 3, 2007, pp. 251-273.
4. M. Miscevic, P. Lavieille and B. Piaud, "Numerical Study of Convective Flow with Condensation of a Pure Fluid in Capillary Regime," International Journal of Heat and Mass Transfer, Vol. 52, No. 21-22, 2009, pp. 5130-5140.
5. G. B. Ribeiro, J. R. Barbosa and A. T. Prata, "Performance of Microchannel Condensers with Metal Foams on the Air-Side: Application in Small-Scale Refrigeration Systems," Applied Thermal Engineering, Vol. 36, 2012, pp. 152-160.
6. G. El Achkar, P. Lavieille and M. Miscevic, "Loop Heat Pipe and Capillary Pumped Loop Design: About Heat Transfer in the Isolated Bubbles Zone of Condensers," Applied Thermal Engineering, Vol. 33-34, 2012, pp. 253- 257.

7. A. Odaymet, H. Louahlia-Gualous, "Experimental Study of Slug Flow for Condensation in a Single Square Microchannel," Experimental Thermal and Fluid Science, Vol. 38, 2012, pp. 1-13.

8. H. Louahlia-Gualous and B. Mecheri, "Unsteady Steam Condensation Flow Pattern inside a Miniature Tube," Applied Thermal Engineering, Vol. 27, 2007, pp. 1225- 1235.

9. A. Odaymet and H. Louahlia-Gualous, "Experimental Study of Slug Flow for Condensation in a Single Square Microchannel," Experimental Thermal and Fluid Science Vol. 38, 2012, pp. 1-13.

10. X. J. Quan, P. Cheng and H. Y. Wu, "Transition from Annular Flow to Plug/Slug Flow in Condensation of Steam in Microchannels," International Journal of Heat and Mass Transfer, Vol. 51, No. 3-4, 2008, pp. 707-716.

11. T. S. Zhao and Q. Liao, "Theoretical Analysis of Film Condensation Heat Transfer inside Vertical Mini Triangular Channels," International Journal of Heat and Mass Transfer, Vol. 45, 2002, pp. 2829-2842.

12. H. Y. Wu and P. Cheng, "Condensation Flow Patterns in Silicon Microchannels," International Journal of Heat and Mass Transfer, Vol. 48, No. 11, 2005, pp. 2186-2197.

13. H. S. Wang, J. W. Rose and H. Honda, "A Theoretical Model of Film Condensation in Square Section Horizontal Microchannels," Chemical Engineering Research and Design, Vol. 82, No. A4, 2004, pp. 430-434.

14. H. S. Wang and J. W. Rose, "Theory of Heat Transfer during Condensation in Microchannels," International Journal of Heat and Mass Transfer, Vol. 54, No. 11-12, 2011, pp. 2525-2534.

15. A. Agarwal., "Heat Transfer and Pressure Drop during Condensation of Refrigerants in Microchannels," Ph.D. Dissertation, Georgia Institute of Technology, 2006.

16. J. Hu and C. Chao, "An Experimental Study of the Fluid Flow and Heat Transfer Characteristics in Micro-Condensers with Slug-Bubbly Flow," International Journal of Refrigeration, Vol. 30, 2007, pp. 1309-1318.

17. Y. Chen, M. Shi, P. Cheng and G. Peterson, "Condensation in Microchannels," Nanoscale and Microscale Thermophysical Engineering, Vol. 12, 2008, pp. 1-27.

18. Y. Chen, J. Wu, M. Shi and G. Peterson, "Numerical Simulation for Steady Annular Condensation Flow in Triangular Microchannels," International Communications in Heat and Mass Transfer, Vol. 35, No. 7, 2008, pp. 805-809.

19. M. M. Shah, "A General Correlation for Heat Transfer during Film Condensation Inside Pipes," International Journal of Heat and Mass Transfer, Vol. 2, 1979, pp. 547-556.

20. W. M .Kays and R. J. Moffat, "The Behavior of Transpired Turbulent Boundary Layers," In: Studies in Convection: Theory, Measurement and Application, Academic Press, London, 1975.

21. Y. Chen, X. Li, J. Wu and M. Shi, "One Dimensional Numerical Simulation for Steady Annular Condensation Flow in Rectangular Microchannels," Heat Mass Transfer, Vol. 46, No. 1, 2009, pp. 75-82.

22. M. K. Dobson, J. C. Chato, D. K. Hinde and S. P. Wang, "Experimental Evaluation of Internal Condensation of Refrigerants R-12 and R-134a," Proceedings of the ASHRAE Winter Meeting, New Orleans, 23-26 January 1994, pp. 744-754.

23. S. Koyama, K. Kuwahara and K. Nakashit, "Condensation of Refrigerant in a Multi-Port Channel," First International Conference on Microchannels and Minichannels, Rochester, 24-25 April 2003, pp. 193-205.

24. D. P. Traviss and W. M, Rohsenow, "Flow Regimes in Horizontal Two-Phase Flow with Condensation," ASHRAE Transactions, Vol. 79, 1973, pp. 31-39.

25. W. W. Akers, H. A. Deans and O. K. Crosser, "Condensation Heat Transfer within Horizontal Tubes," Chemical Engineering Progress Symposium Series, Vol. 55, No. 29, 1959, pp. 171-176.

26. W. W. Wang, T. D. Radcliff and R. N. A. Christensen, "Condensation Heat Transfer Correlation for MillimeterScale Tubing with Flow Regime Transition," Experimental Thermal and Fluid Science., Vol. 26, No. 5, 2002, pp. 473-485.

3
CHAPTER

FEM APPROACH FOR TRANSIENT HEAT TRANSFER IN HUMAN EYE

Gokul K. C.*, Dil Bahadur Gurung, Pushpa Raj Adhikary

Department of Natural Sciences (Mathematics), School of Science, Kathmandu University, Kavre, Nepal

ABSTRACT

In this paper, a bio-heat transfer model of temperature distribution in human eye is discussed using appropriate boundary conditions for cornea and sclera. Variational finite element method with Crank-Nicolson scheme is used to calculate the transient temperature distribution in normal human eye. The temperature with and without the effect of blood perfusion and metabolism on retina is simulated and compared for various ambient temperatures, evaporation rates and lens thermal conductivities. The obtained results are compared with experimental results and past results found in literatures. The results show that the steady state corneal temperature is achieved in around 31 and 45 minute of exposure at ambient temperatures 10°C and 50°C respectively. Steady state eye temperature is achieved earlier at higher evaporation rate. Similar result is achieved for higher lens thermal conductivity and also for lower ambient temperature.

Keywords: Human Eye; Bio-Heat Transfer; FEM; Numerical Simulations

1. INTRODUCTION

In a human body the internal body temperature almost remains constant despite the fluctuation of environmental temperatures up to certain limits. The main organ that keeps core temperature constant is dermal part [1]. There is no skin layer to keep the core temperature constant in case of human eye. The skin layer (eyelid) covers eye surface (cornea) for 3 seconds in a minute (in an average). But for 57 seconds in a minute,ocular surface (cornea) has to manage thermal stress of an environment. The human eye is relatively a small and complex organ, consists of several sub domains with different material properties and having complex geometry.

The calculation of the temperature distribution in human eye when it is heated or cooled is an important aspect of the development of infrared and radiofrequency safety guidelines and for hyperthermia and thermo-therapy treatments of various ocular diseases [2]. The severity of the physiological effect produced by small temperature increases can cause eyesight to worsen. Actually, a small temperature increase in the eye of 3°C - 5°C leads to induce cataracts formation [3]. Some researchers believe that thermal effect can induce cataracts; other believes that it is the result of other biological and genetic issues. One of the early theory suggested that heat exchanges within the anterior eye caused the cataract. Verhoeff and Bell argued that cataract formed on the posterior surface of the lens because the anterior surface was cooled by circulation of the aqueous humor and so the cornea was air cooled [4]. Investigation in Germany showed that the cataracts were due to the raised temperature induced indirectly through heat absorbed by the iris, where a rich blood supply would be consistent with a high degree of heating. At the same time Salil noticed the rise in cataracts one year after a very hot, dry summer in Iowa, highlighting likely environmental causes of cataracts [4].

Due to convective heat transport of the blood vessels; the blood picks up energy from hot areas and deposits this at cooler area or vice versa. The difficulty of modeling in the eye is due to the impact of blood flow on the heat transfer; however, incorporating the impact of blood flow in heat transport calculation is very important [2]. The temperature inside the human body depends on the degree of temperature, duration of exposure and the environmental conditions which cause heat gain/loss from tissues [1]. Hence, blood flow and time are the main factors that affect temperature distribution in human eye.

Lagendijk [5] used a finite difference method to calculate the temperature distribution in human and rabbit eyes during hyperthermia treatment. The heat transport from the sclera to the surrounding anatomy is described by a single heat transfer coefficient which includes the impact of blood flow in choroid and sclera. Scott [6] utilized finite element method to obtain the temperature profile based on heat conduction using various heat transfer coefficients given by Lagendijk. Amara [7] presented a numerical thermal model of laser-ocular media interaction. Ng and Ooi [8] studied the effect of aqueous humor hydrodynamics on heat transfer within human eye by neglecting blood perfusion and metabolism. They neglected the effects because they assumed that perfusion in iris/cilliary is sufficient to maintain their temperature at

37°C. Li et al. [9] studied the bio-heat transfer in the human eye, neglecting the effect of perfusion and metabolism, using 3D alpha finite element method. They assumed the contribution due to perfusion and metabolism is very small because they occur only on small part in the human eye. Cvetkovic et al. [10] developed a thermal model and studied effects of pulsed laser in human eye. Narasimhan et al. [11] developed transient model to study heat transfer in human eye undergoing laser surgery. Flyckt et al. [2] studied the impact of choroidal blood flow and scleral convection heat transfer coefficient in human eye.

All the previously developed models have neglected the effects of blood perfusion and metabolism on retina/iris/cilliary body. The significance of blood perfusion and metabolism on the temperature distribution in the eye is debatable, since they take place only on retina, choroid, iris and cilliary body, which constitute very small part of human eye. The blood flow in the iris/sclera part plays significant role to adjust eye temperature with the rest of the body [3]. In [8] it is clearly mentioned that "Due to lack of literature data, the perfusion term is neglected. The effect of this assumption on the accuracy of the model however remaining unknown". The retina has perhaps the highest oxygen consumption rate (metabolism) of any tissue in the body [12]. Hence it is necessary to investigate the effects of blood perfusion and metabolism in order to obtain a more accurate result. The objective of this paper is to determine the transient temperature distribution in human eye. Hence, this model provides the results for transient temperature distribution including the effect of blood perfusion and metabolism.

2. MODEL FORMULATION

The eye is assumed a perfectly bonded solid structure with each component homogeneous. The eye is considered having six major components: cornea, aqueous humor, lens, vitreous humor, retina, and sclera. As sketched on Figure 1, the thickness of cornea, aqueous humor, lens, vitreous humor, retina and sclera have been considered as $l_1, l_2 - l_1, l_3 - l_2, l_4 - l_3, l_5 - l_4$ and $l_6 - l_5$ respectively. Similarly, $T_0, T_1, T_2, T_3, T_4, T_5$ and $T_6 = T_b$ (body core temperature) are the nodal temperatures at a distances $x = 0, x = l_1, x = l_2, x = l_3, x = l_4, x = l_5$ and $x = l_6$.

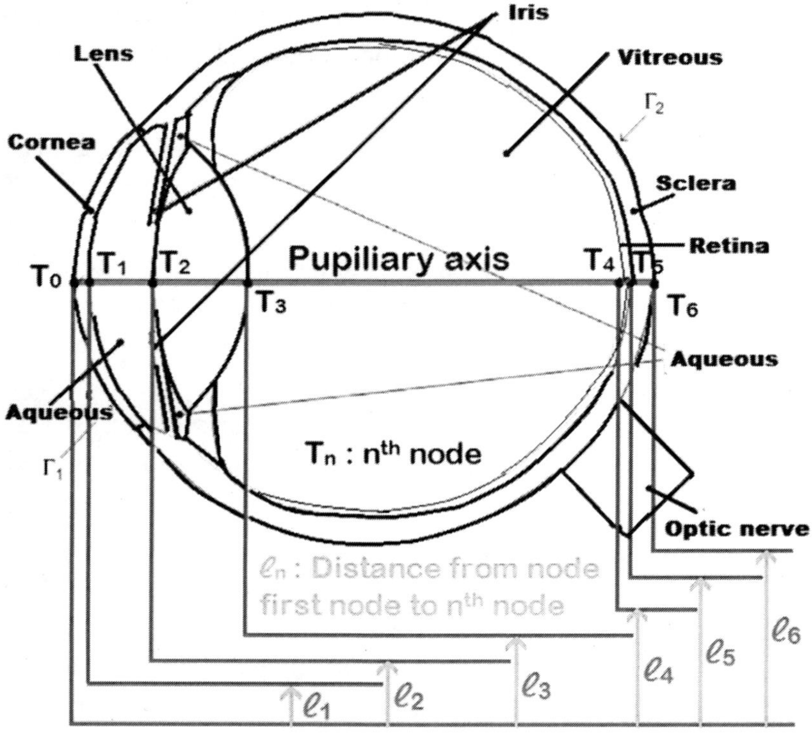

Figure 1. Finite element sketch of human eye.

A one dimensional finite element human eye model has been developed to simulate its thermal unsteady state conditions. The governing differential equation used for heat flow in the human eye due to Pennes [13] is given by:

$$\rho c \frac{\partial T}{\partial t} = \frac{\partial}{\partial x}\left(k \frac{\partial T}{\partial x}\right) + \omega \rho_b c_b \left(T_b - T\right) + Q_m \tag{1}$$

where, ρ = tissue density (kg/m^3), c = tissue specific heat (J/kg°C), ρ_b = blood density (kg/m^3), c_b = blood specific heat (J/kg°C), k = tissue thermal conductivity (W/m°C), ω = volumetric blood perfusion rate per unit volume (s^{-1}), T_b = blood temperature (°C), T = tissue temperature (°C), t = time (s), Q_m = heat generation due to metabolism (W/m^3).

The three terms on the right-hand side of bio-heat Equation (1) are blood conduction, blood perfusion, and metabolism. In this model, the effect of blood perfusion and metabolism is analyzed only on retina.

Boundary conditions for the system can be defined as follows [9,14]:

1) In the back of the eye, heat is transferred from blood via ophthalmic vessels to the sclera:

$$\Gamma_2 : -k_s \frac{\partial T}{\partial \eta} = h_b \left(T - T_b\right) \tag{2}$$

where η is the normal direction to the surface boundary, k_s is the thermal conductivity of sclera, h_b is the heat transfer coefficient between blood and eye (W/m^2 °C), and T_b is blood temperature (°C).

2) At the cornea, heat loss from the eye occurs through convection, radiation, and tear evaporation:

$$\Gamma_1 : -k_c \frac{\partial T}{\partial \eta} = h_\infty \left(T - T_\infty\right) + \sigma \varepsilon \left(T^4 - T_\infty^4\right) + E \tag{3}$$

where $h_{¥}$ represents the convection heat transfer coefficient between the cornea and ambient environment (W/m^2 °C), $T_{¥}$ is the ambient room temperature (°C), σ is the Stefan Boltzmann constant

(5.67×10^{-8} W/m^2°C^4), ε is the emissivity of the cornea, and E is evaporative heat loss (W/m^2).

The nonlinear radiation term in the boundary condition (3) is treated by using simple iterative procedure as follows:

$$-k_c \frac{\partial T_1}{\partial n} = \left[h_\infty + \sigma\varepsilon\left(T_1 + T_\infty\right)\left(T_1^2 + T_\infty^2\right)\right]\left(T_1 - T_\infty\right) + E \tag{4}$$

$$-k_c \frac{\partial T_1^m}{\partial n} = h_{cr}\left(T_1^m - T_\infty\right) + E \tag{5}$$

where,

$$h_{cr} = h_\infty + \sigma\varepsilon\left(T_1^{m-1} + T_\infty\right)\left(\left(T_1^{m-1}\right)^2 + T_\infty^2\right) \tag{6}$$

$$h_{\mathrm{cr}} = h_{\mathrm{convection}} + h_{\mathrm{radiation}}$$

where T_1^m are temperature sequences for $m = 1, 2, 3, \cdots$ and T_1^0 represents an initial guess of temperature.

The iteration is completed when the convergent condition is satisfied:

$$\left|T_1^m - T_1^{m-1}\right| < \delta \tag{7}$$

where δ is iteration tolerance.

The inner body core temperature T_b is assumed to be 37°C. Therefore, the initial boundary condition is

$$T_b = 37□\mathrm{C} \tag{8}$$

The partial differential Equation (1) together with boundary conditions (2) and (5) in one dimensional variational form is:

$$I = \frac{1}{2}\int_L \left[K\left(\frac{dT}{dx}\right)^2 + \omega\rho_b c_b (T_b - T)^2 - 2Q_m T + \rho c \frac{\partial T^2}{\partial t} \right]$$
$$\cdot dx + \frac{1}{2} h_b (T - T_b)^2 + \frac{1}{2} h_{cr} (T - T_\infty)^2 + ET \quad (9)$$

We write I separately for the six layers in the following equation:

$$I = \sum_{i=1}^{6} I_i \quad (10)$$

Further, to optimize I, we differentiate I partially with respect to T_i and equating to zero, we get,

$$\frac{\partial I}{\partial T_i} = 0, \ i = 0, 1, 2, \cdots, 5 \quad (11)$$

The system of Equations (11) can be written in matrix form

$$[C]\{\dot{T}\} + [K]\{T\} = \{R\} \quad (12)$$

where $\{\dot{T}\} = \left\{\frac{\partial T_i}{\partial t}\right\}$, $\{T\} = \{T_i\}$ and $\{R\} = \{R_i\}, i = 0, 1, 2, \cdots, 5$ are 6×1 vectors and [C] and [K] are 6×6 matrices called capacity and conductivity matrices respectively.

Now we apply Crank-Nicolson method to solve the system (12) with respect to time, we get the following relation

$$\left(\frac{1}{\Delta t}[C] + \frac{1}{2}[K]\right)\{T\}_{n+1} = \left(\frac{1}{\Delta t}[C] - \frac{1}{2}[K]\right)\{T\}_n + \{R\} \quad (13)$$

where Δt is time interval.

The temperature increases from outer surface of cornea towards eye core when ambient temperature is less than 37°C and vice versa. Hence, we consider the temperature increases/decreases in linear order towards body core with regard to thickness. For initial nodal temperatures $\{T\}_0$ at time $t = 0$, we assume the following initial condition

$$T\left(x = l_i, t = 0\right) = T\left(0,0\right) + rl_i,\ i = 1,2,3,\cdots,6 \tag{14}$$

where $T(0,0) =$ 20°C and $r =$ constant to be determined. The Equation (13) is repeatedly solved to get the required nodal temperatures.

3. RESULTS AND DISCUSSION

To solve equation (13), the following values of parameters are considered [14]: h_∞ = 10 W/m^2 °C, h_b = 65 W/m^2 °C, r_b = 1060 Kg/m^3, c_b = 3594 J/Kg °C, K_1 = 0.580 W/m °C, K_2 = 0.578 W/m °C, K_3 = 0.4 W/m °C, K_4 = 0.603 W/m °C, K_5 = 0.565 W/m °C, K_6 = 1.0042 W/m °C, l_1 = 0.0005 m, l_2 = 0.00354 m, l_3 = 0.00754 m, l_4 = 0.02355 m, l_5 = 0.02405 m, l_6 = 0.02510 m, w = 0.012 s^{-1}, Q_m = 1000 W/m^3, $T_1^0 = 0$, and $\delta = 0.0005$.

The numerical calculation for unsteady state temperature distribution is carried out for different parts of human eye with and without taking blood perfusion and metabolism on the retina. The effects of different ambient temperatures, tear evaporation rates and lens thermal conductivities are studied and compared. The results obtained are compared with experimental and other results found in literatures. The overall thermal behavior of human eye is observed for 3600 seconds using one second time step size.

3.1. Effect of Blood Perfusion and Metabolism

The transient temperature distribution of several parts of human eye for $t = 5000s$ is presented in Figures 2 and 3 below. The parameter values $E = 40\ W/m^2$, K_3 = 0.4 W/m °C and $T_\infty = 10^\circ C$ and $50^\circ C$ respectively are used for analysis.

It can be seen from Figures 2 and 3 that the anterior and posterior part of cornea, aqueous and lens temperatures begin to stabilize around 1870 seconds (approximately 31.17 minutes) and 2740 seconds(approximately 45.67 minutes) respectively. In human eye, heat gain occurs through conduction, perfusion, metabolism, blinking, tear flow, evaporation, and convection but heat loss occurs only through conduction, evaporation, convection and radiation. More factors are involved in heating eye components than cooling. Hence, eye is more vulnerable when it is exposed to high temperatures (high ambient temperatures, hyperthermia treatment, laser surgery etc.) than low (low ambient temperatures, cryosurgery treatment etc). The temperature difference obtained between anterior and posterior parts of cornea are 2.2°C and 0.65°C respectively at $T_\infty = 10^\circ C$ and T_∞ = 50°C. As well the temperature differences between anterior and posterior parts of lens are 0.2°C and 0.05°C.

The corneal temperature distribution with and without blood perfusion and metabolism in retina at T_∞ = 10°C and T_∞ = 50°C are shown in Figures 4 and 5 below.

It can be observed from Figures 4 and 5 that when corneal surface temperature is attained steady state in both cases, the temperature differences with and without blood perfusion and metabolism obtained are 0.23°C and

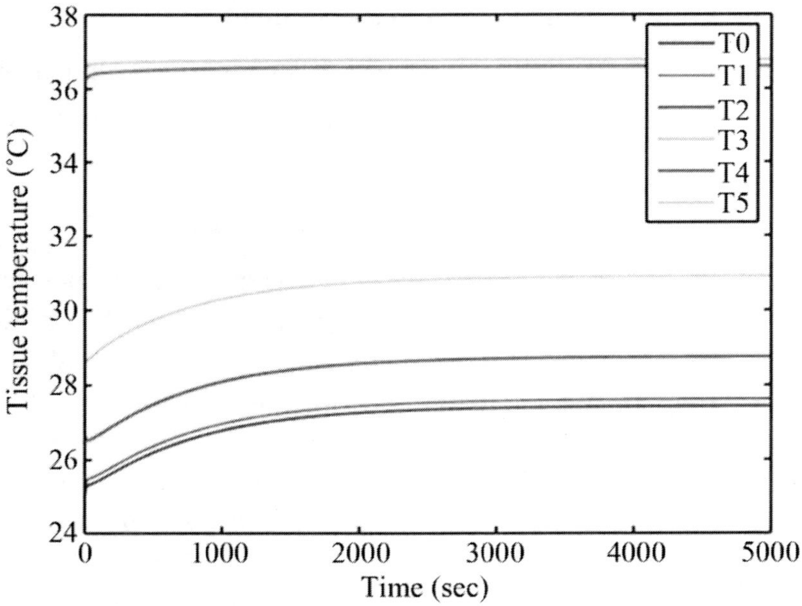

Figure 2. Temperature variation at $T_\infty = 10$°C.

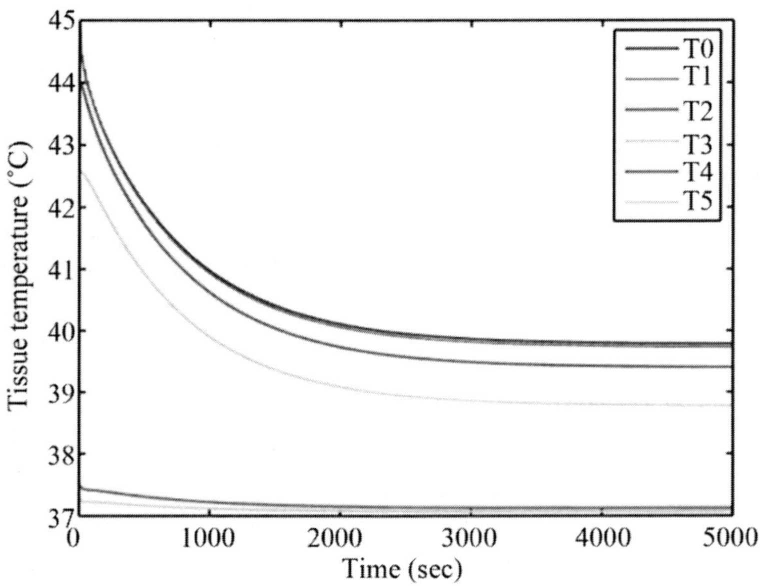

Figure 3. Temperature variation at $T_\infty = 50$°C.

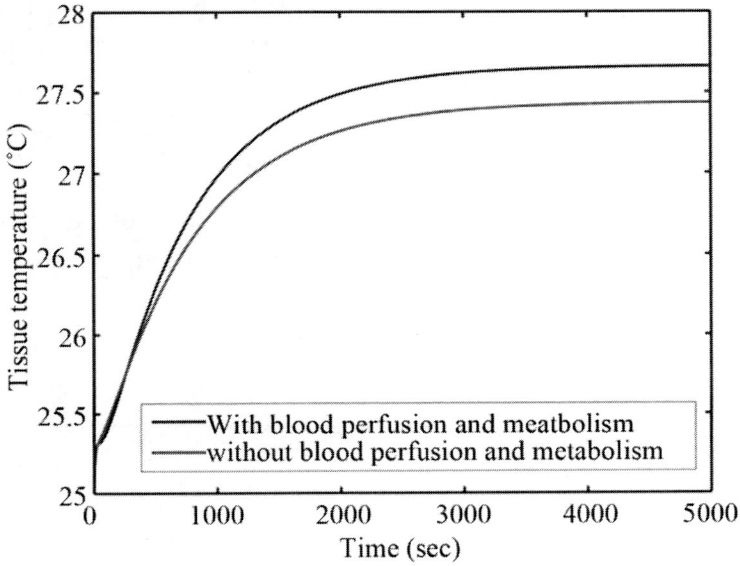

Figure 4. Temperature variation at $T_\infty = 10°C$.

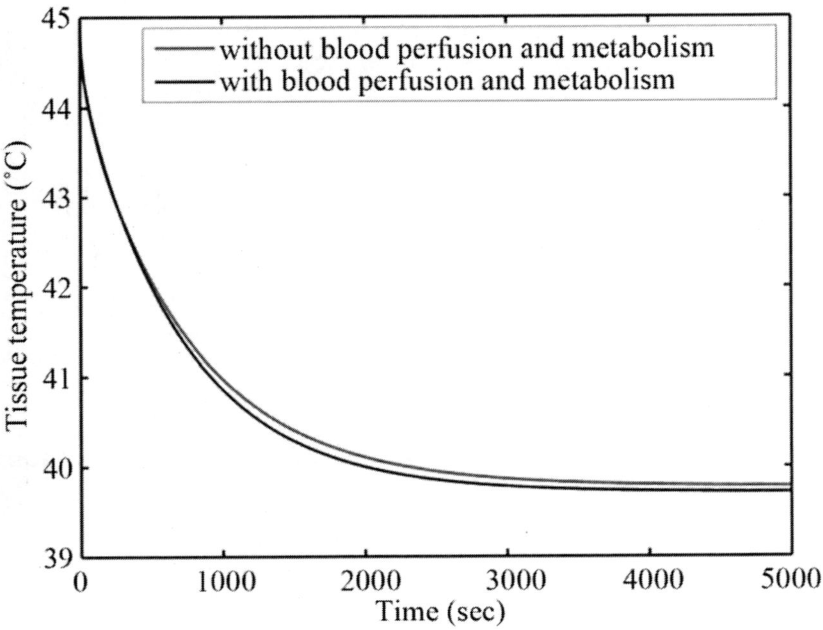

Figure 5. Temperature variation at $T_\infty = 50°C$.

0.06°C respectively. Steady state corneal surface temperature is reached earlier in case of having blood perfusion and metabolism in retina. Accordingly, steady state corneal surface temperature is reached earlier at T_∞ = 10°C compared to that at T_∞ = 50°C. This is due to better heating mechanism than cooling in eye.

For further investigation, we now put the effect with blood perfusion and metabolism as Case I and without them as Case II.

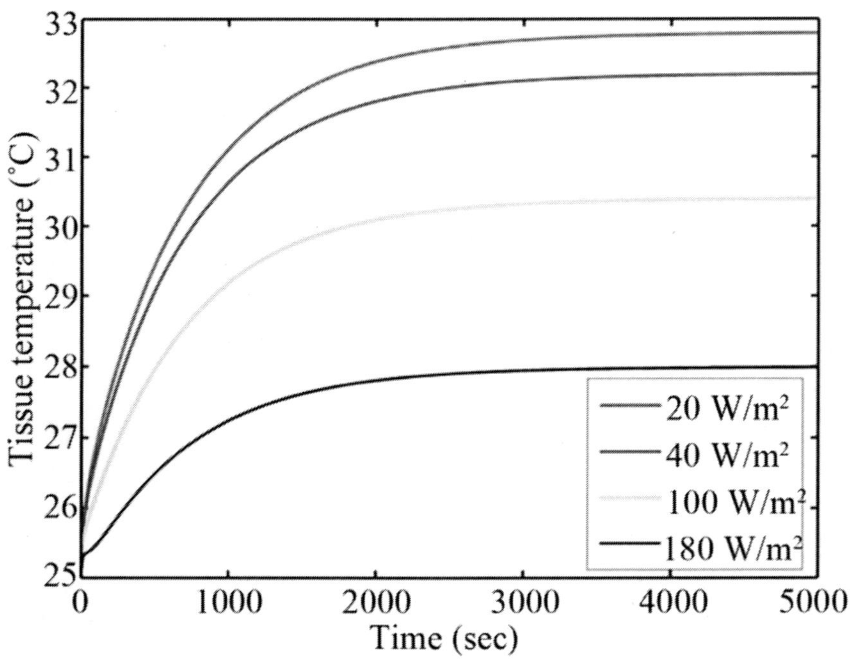

Figure 6. Corneal surface temperature for different evaporation rates (Case I).

3.2. Effect of Tear Evaporation

The corneal surface contains a three-layered structure—a mucoid layer, a thick aqueous layer and thin oily layer. The function of oily layer is to prevent evaporation of tear from the corneal surface. When the oily layer is destroyed, the evaporation rate increases dramatically [15]. Four sets of data values for E equals 20 W/m², 40 W/m², 100 W/m², 180 W/m² are used in this investigation. The temperature variations for different evaporation rates at $T_{\infty} = 25^{\circ}C$, $K_3 = 0.4\ W/m^{\circ}C$ are shown in Figures 6 and 7.

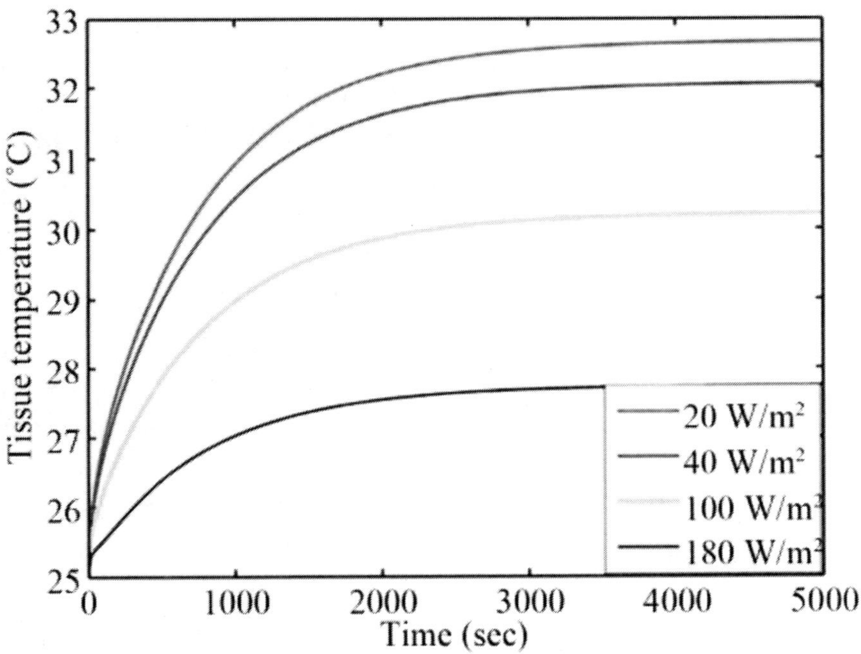

Figure 7. Corneal surface temperature for different evaporation rates (Case II).

For different evaporation rates of 20 W/m^2, 40 W/m^2, 100 W/m^2, 180 W/m^2, the corneal surface temperature differences between Case I and Case II are obtained as 0.10°C, 0.12°C, 0.18°C, and 0.27°C respectively. The differences show that when evaporation rate increases the nodal temperature decreases in both cases but decreasing rate is slow in Case I than Case II. It means, steady state corneal surface temperature is reached earlier in Case I than Case II. This is due to the effect of perfusion and metabolism on retina. The corneal surface temperature is dropped in by 4.78°C and 4.95°C in Cases I and II respectively, when evaporation rate is reached from 20 W/m^2 to 180 W/m^2. Also, steady state corneal temperature is achieved earlier at E = 180 W/m^2 than at 20 W/m^2. The sudden decrease in corneal temperature at E = 180 W/m^2 increases the temperature difference between cornea and aqueous. It is because the greater the temperature difference between two surfaces, the faster is the rate of transfer of thermal energy. Hence steady state corneal temperature is reached earlier at higher evaporation rate.

3.3. Effect of Ambient Temperatures

Heat losses occur due to convection, radiation and tear evaporation at cornea. This loss is strongly related to ambient temperature. In addition, ambient temperature is one of the factors affecting the amount of tear in the eyes [9]. Four sets of ambient temperatures 10°C, 25°C, 40°C, and 50°C are considered for analysis. The numerical results are presented in Figures 8 and 9.

It can be seen from Figures 8 and 9 that, increase in ambient temperature from 10°C to 50°C increases corneal temperature from 27.69°C to 39.69°C in Case I and from 27.41°C to 39.78°C in Case II. It shows that increase in ambient temperature increases eye temperatures lower in case I compared to Case II and vice versa. This is due to the cooling effect of perfusion at retina. Also, the steady state corneal temperature is reached earlier at T_∞ = 10°C than at T_∞ =

50°C. This is because; more factors are involved in heating eye components than cooling.

3.4. Effect of Lens Thermal Conductivities

It is well known that the water content of lens decreases as age increases. Decrease in water level of lens increases its hardness. This process changes the thermal conductivity of lens due to age [14]. Four sets of values 0.21 W/m °C, 0.30 W/m °C, 0.40 W/m °C and 0.54 W/m °C are chosen for comparison. The numerical results are presented in Figures 10 and 11.

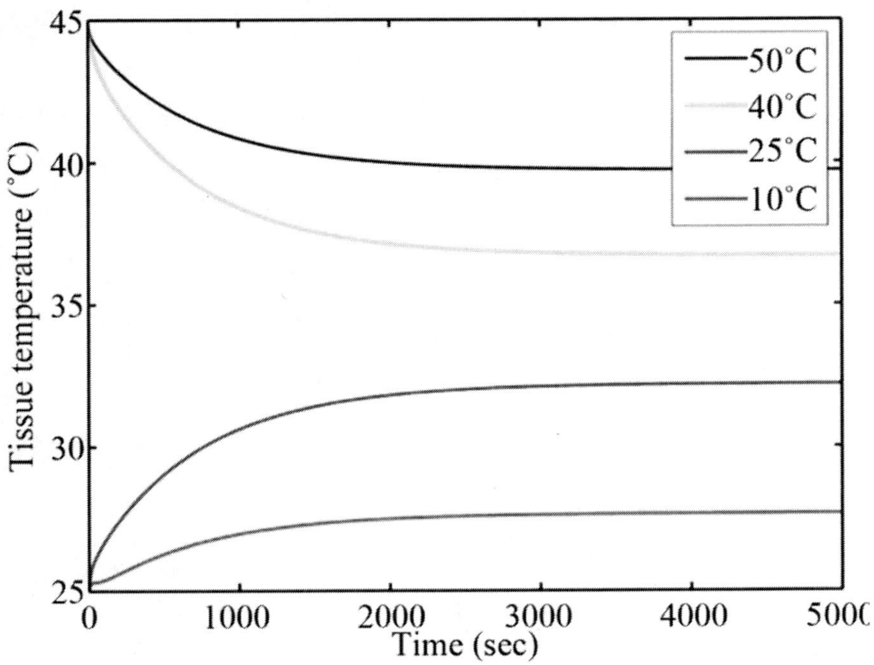

Figure 8. Corneal surface temperature for different ambient temperatures (Case I).

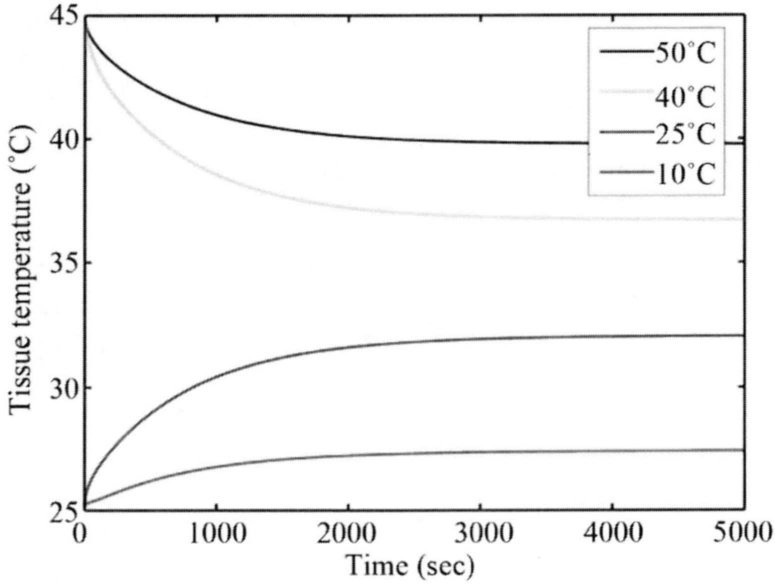

Figure 9. Corneal surface temperature for different ambient temperatures (Case II).

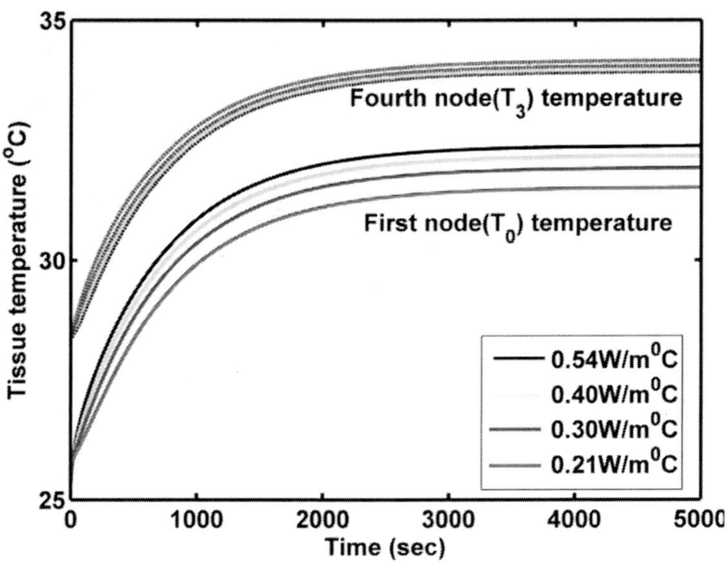

Figure 10. Anterior corneal and posterior lens temperature for different Lens thermal conductivities (Case I).

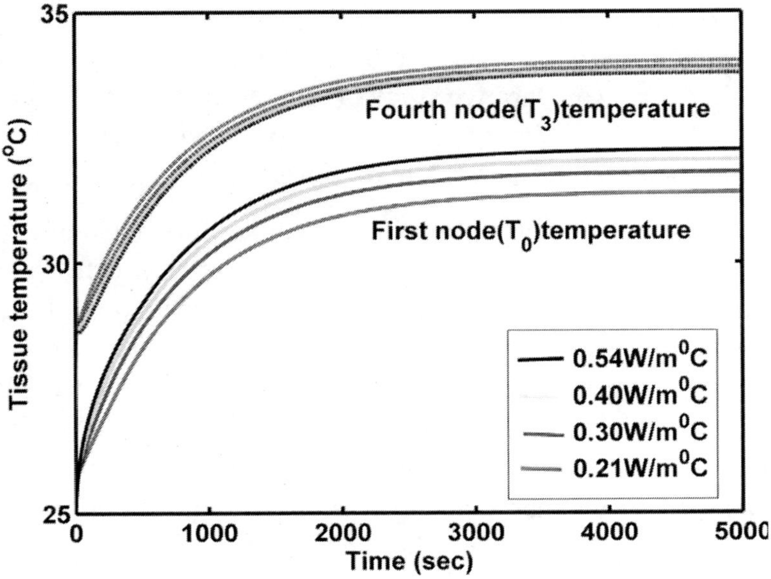

Figure 11. Anterior corneal and posterior lens temperature for different Lens thermal conductivities (Case II).

It is found that, increase in lens thermal conductivity from 0.21 W/m °C to 0.54 W/m °C increases corneal temperature by 0.87°C in Case I and by 0.85°C in Case II but decreases posterior lens temperature by 0.23°C in Case I and by 0.25°C in Case II. More heat transfer occurs via conduction from the posterior region to anterior region, when thermal conductivity of lens increases. As a result corneal surface temperature is increased. The steady state corneal temperature is achieved earlier at K_3 = 0.54 W/m °C than at 0.21 W/m °C but the process is reversed in case of posterior part of lens. Similarly, steady state eye temperature is reached earlier in Case I compared to Case II due to the effect of perfusion and metabolism on retina.

4. CONCLUSIONS

In this model, a comparative study of temperature distribution with and without considering the effect of blood perfusion and metabolism on retinal part of human eye is

presented. The steady state corneal temperature is found to be 32.17°C in Case I and 32.05°C in Case II respectively. Many authors [1,3,10,13,15] reported similar temperature distribution along the eye pupillary axis. The reported values range from 30.92°C to 33.7°C on corneal surface. Hence, our steady state results are in accordance with past results. In earlier studies [1,3,10,13,15], the effect of blood perfusion and metabolism is assumed negligible on retinal part. We are able to show that blood perfusion and metabolism play an important role in maintaining the eye temperature.

The surface temperature of the cornea was measured using bolometer by Mepastone [16], the mean temperature variation on the cornea surface was obtained to be 0.8°C under 33.2°C - 36°C. Kessel et al. [17] found that steady state corneal temperature was achieved between 33°C to 35°C, when ambient temperature was increased. Also they found that 20°C increase in ambient temperature, from 2°C to 22°C, was required to increase corneal temperature by 3°C. In our model, the steady state corneal temperature is achieved between 32.17°C to 35.58°C, when ambient temperature is increased from 20°C to 40°C. Also, a 20°C increase in ambient temperature is required to increase corneal temperature by 3.38°C. Hence, the temperature obtained from our model agrees with the experimental results obtained by Mepastone [16] and Kessel et al. [17]. The slight differences whatsoever obtained may be due to consideration of various parameter values at the layers of human eye.

This model may help to understand the thermal behavior of eye, which can be very crucial in ocular diseases like corneal pain, presbyopia, cataracts etc. This may also prove to be valuable in several eye therapies and surgeries like hyperthermia using microwave, heating due to laser etc. Thus, the model could be useful for the researchers to study the effects of heat flux inside the human eye and medical scientists to improve the diagnosis and treatment.

5. ACKNOWLEDGEMENTS

The authors would like to thank University Grants commission Nepal for providing financial support to conduct this research work.

REFERENCES

1. D. B. Gurung and V. P. Saxena, "Transient Temperature Distribution in Human Dermal Part with Protective Layer at Low Atmospheric Temperature," International Journal of Biomathematics, Vol. 3, No. 4, 2010, pp. 439-451.
2. V. M. M. Flycket, B. W. Roaymakers and J. J. W. Lagendijk, "Modeling the Impact of Blood Flow on the Temperature Distribution in the Human Eye and the Orbit: Fixed Heat Transfer Coefficient versus the Pennesbioheat Model versus Discrete Blood Vessels," Physics in Medicine and Biology, Vol. 51, No. 19, 2006, pp. 5007-5021.
3. T. Wessapan and P. Rattanadecho, "Specific Absorbtion Rate and Temperature Increase in Human Eye Subjected to Electromagnetic Fields at 900 MHZ," Transactions of the ASME-C-Journal of Heat Transfer, Vol. 134, No. 9, 2012, Article ID: 091101.
4. J. Voke, "Occupational Vision Hazards: Infrared Radiation and the Eye," Optometry Today, Vol. 48, No. 17, 2008, pp. 40-43.

5. J. J. W. Lagendijk, "A Mathematical Model to Calculate Temperature Distributions in Human and Rabbit Eyes during Hyperthermia Treatment," Physics in Medicine and Biology, Vol. 27, No. 11, 1982, pp. 1301-1311.

6. J. A. Scott, "A Finite Element Model of Heat Transport in the Human Eye," Physics in Medicine and Biology, Vol. 33, No. 2, 1988, pp. 227-241.

7. E. H. Amara, "Numerical Investigations on Thermal Effects of Laser-Ocular Media Interaction," International Journal of Heat and Mass Transfer, Vol. 38, No. 13, 1995, pp. 2479-2488.

8. E. Y. K. Ng and E. H. Ooi, "FEM Simulation of the Eye Structure with Bio-Heat Analysis," Computer Methods and Programs in Biomedicine, Vol. 82, No. 3, 2006, pp. 268-276.

9. E. Li, G. R. Liu, V. Tan and Z. C. He, "Modeling and Simulation of Bioheat Transfer in the Human Eye Using the 3D Alpha Finite Element Method (aFEM)," International Journal for Numerical Methods in Biomedical Engineering, Vol. 26, No. 8, 2010, pp. 955-976.

10. M. Cvetkovic, D. Poljak and A. Peratta, "FETD Computation of the Temperature Distribution Induced into a Human Eye by a Pulsed Laser," Progress in Electromagnetics Research, Vol. 120, 2011, pp. 403-421.

11. A. Narasimhan, K. K. Jha and L. Gopal, "Transient Simulations of Heat Transfer in Human Eye Undergoing Laser Surgery," International Journal of Heat and Mass Transfer, Vol. 53, No. 1, 2009, pp. 482-490.

12. B. Anderson, "Ocular Effects of Changes in Oxygen and Carbon Dioxide Tension," Transactions of the American Ophthalmological Society, Vol. 66, No. 1, 1968, pp. 423- 474.

13. H. H. Pennes, "Analysis of Tissue and Arterial Blood Temperatures in the Resting Human Forearm," Journal of Applied Physiology, Vol. 85, No. 1, 1998, pp. 5-34.

14. K. C. Gokul, D. B. Gurung and P. R. Adhikary, "Effect of Blood Perfusion and Metabolism in Temperature Distribution in Human Eye," Advances in Applied Mathematical Biosciences, Vol. 4, No. 1, 2013, pp. 13-23.

15. H. Wang and Q. H. Qin, "FE Approach with Green's Function as Internal Trial Function for Simulating Bioheat Transfer in the Human Eye," Archives of Mechanics, Vol. 62, No. 6, 2010, pp. 493-510.

16. R. Mepastone, "Determinants of Corneal Temperature," British Journal of Ophthalmology, Vol. 52, No. 10, 1968, 729-741.

17. L. Kessel, L. Johnson, H. Aridsson and M. Larsen, "The Relationship between Body and Ambient Temperature and Corneal Temperature," Investigative ophthalmology & Visual Science, Vol. 51, No. 12, 2010, pp. 6593-6597.

4

CHAPTER

MEMS-BASED BOILER OPERATION FROM LOW TEMPERATURE HEAT TRANSFER AND THERMAL SCAVENGING

Suvhashis Thapa, Emmanuel Ogbonnaya, Christopher Champagne and Leland Weiss *

Department of Mechanical Engineering, Louisiana Tech University, PO Box 10348 Ruston, LA

ABSTRACT

Increasing world-wide energy use and growing population growth presents a critical need for enhanced energy efficiency and sustainability. One method to address this issue is via waste heat scavenging. In this approach, thermal energy that is normally expelled to the environment is transferred to a secondary device to produce useful power output. This paper investigates a novel MEMS-based boiler designed to operate as part of a small-scale energy scavenging system. For the first time, fabrication and operation of the boiler is presented. Boiler operation is based on capillary action that drives working fluid from surrounding reservoirs across a heated surface. Pressure is generated as working fluid transitions from liquid to vapor in an integrated steamdome. In a full system application, the steam can be made available to other MEMS-based devices to drive final power output. Capillary channels are formed from silicon substrates with 100 μm widths. Varying depths are studied that range from 57 to 170 μm. Operation of the boiler shows increasing flow-rates with increasing capillary channel depths. Maximum fluid mass transfer rates are 12.26 mg/s from 170 μm channels, an increase of 28% over 57 μm channel devices. Maximum pressures achieved during operation are 229 Pa.

Keywords: MEMS; micro heat engine; phase change; boiler; free piston; scavenging; micro heat transfer

1. INTRODUCTION

Pressing energy concerns and growing rates of worldwide energy consumption present a need for focused research on thermal energy scavenging and power production from renewable sources. In 2010 the United States consumed 98 quadrillion BTUs (British Thermal Units) of energy [1]. The U.S. Department of Energy has further predicted that global energy consumption will increase by 44% compared to 2006 levels [2]. Of particular concern is the operating efficiency of many standard cycles. The Otto cycle, for example, exhibits an approximate efficiency of 35% [3]. The remaining two thirds of initial fuel energy is expelled into the surrounding environment as waste thermal energy from either exhaust or heat exchanger devices. This is of particular concern as transportation sources account for nearly 30% of total energy consumed in the U.S. on an annual basis [1]. Other sectors of energy consumption produce large amounts of waste thermal energy as part of their regular operation as well.

The work in this paper presents investigations of a unique phase-change based steam system designed to capture waste heat from some of the sources noted previously. The use of this normally wasted energy provides an overall efficiency enhancement for the larger process as a whole. The system is designed around small-scale, microfabricated devices that may operate across the low temperature gradients commonly associated with waste thermal energy sources. As an added benefit, the devices developed as part of this waste heat scavenging effort are equally well suited to exploit other sources of environmental thermal energy. These include solar thermal sources [4].

Micro-devices and engines have been under investigation by a wide variety of research groups for a number of years. One of the first engines was a micro steam engine constructed by Sandia National Labs in 1993 [5]. Since then, efforts have expanded to include micro steam turbines operating on the Rankine cycle [6,7]. There has also been interest in micro engines operating on the Brayton and Wankel cycles [8,9,10,11]. One challenge to many of these approaches has been the nature of high speed rotating components when implemented on the micro scale. High speed stability and bearing tolerance are among the engineering and fabrication challenges. Nonetheless, progress continues on these unique devices [7]. Other micro-engine concepts have relied on power output that does not yield rotary motion. Work by Weiss *et al.* included a flexing-expander membrane-based device that outputs power via membrane expansion [12,13,14]. The device consisted of two membranes, separated by a gasket and filled with working fluid that boils at low temperature. When thermal energy was applied, the working fluid expanded causing membrane expansion and power output. This device represented a thermodynamically closed operating cycle, where no working fluid crossed the boundaries established by the device volume. Thermal energy was introduced in discrete applications through the inclusion of a novel thermal switch [15,16].

This present work investigates the operation of a system in an open thermodynamic cycle. The complete system is shown in Figure 1. In its final form, low temperature waste heat (or thermal energy from another source) is consumed by the system boiler where working fluid transitions from liquid to vapor. The resulting pressurized vapor is expanded by a unique Free Piston Expander (FPE) to produce the work output from the system. Expanded working fluid is cooled via heat exchanger and then returned to the boiler for reuse. In this manner, the system represents a modified Rankine Cycle which lacks the typical rotary expansion and high operating temperature.

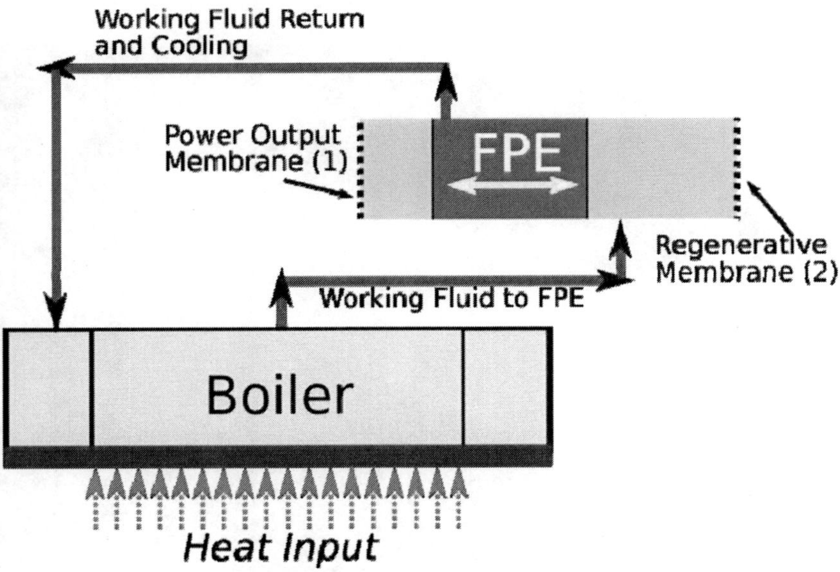

Figure 1. System overview (FPE—Free Piston Expander).

The boiler and FPE are uniquely designed components of this system. The FPE has been modeled and discussed in previously published work [17]. The free piston (noted as "FPE" in Figure 1) slides linearly within a bore based on input pressure from the system boiler. As the piston slides, it expands the working fluid and pressurizes the primary power output membrane. Once fully expanded, the vapor is released to the cooling line and the piston begins its return to the starting position. The return is driven by energy stored within the primary membrane. The secondary, regenerative membrane helps absorb piston energy during the return stroke and aids in the overall efficiency of the system through this unique regenerative capture. The action of the membranes constitutes the system power output and is designed to produce either mechanical or electrical power. Electrical power may be created through the actuation of an external PZT cantilever or similar device. Mechanical power output is achievable through direct membrane mechanical action. Figure 2 shows an early FPE bore.

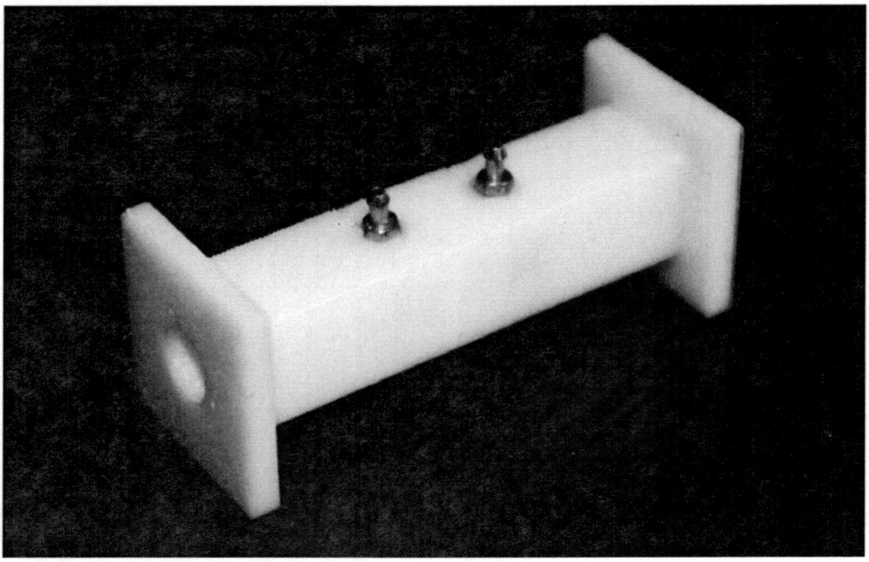

Figure 2. FPE prototype bore.

The boiler is the particular focus of this present work. Like the FPE, the boiler is a unique design for low temperature application. The design focuses on the ability to generate working fluid pressure with minimal moving parts or components common to large-scale steam systems or true Rankine cycle devices on the microscale. This paper represents the first time a boiler of this type has been produced via MEMS techniques and operated at pressurized conditions. Figure 3 illustrates the boiler concept.

Operation of the boiler is dependent on capillary action of the working fluid across the heated lower surface as noted in Figure 3. This capillary action is based on the microfabrication of capillary channels that take working fluid from the surrounding fluid reservoirs, providing a continuous source of working fluid supply to the heated surface. The top of the boiler is enclosed with a steamdome to provide a central collection point for vaporized working fluid. From the steamdome, the pressurized working fluid is available to devices like the FPE in the final assembly of the overall system.

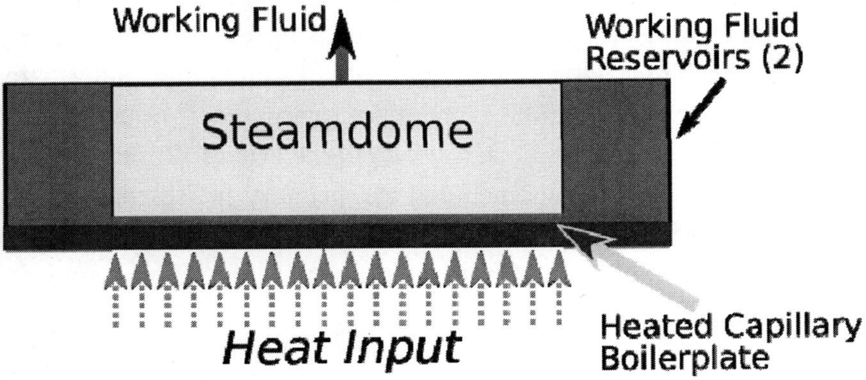

Figure 3. Boiler design.

The overall approach to this low-temperature energy challenge has several advantages including ease of fabrication due to the lack of high speed bearings or rotating parts. Operation of the device at temperatures typical of environmentally scavenged waste heat allows a wide degree of flexibility in material selection as well. The remainder of this work discusses the fabrication and characterization of the boiler. Reported testing considers variations like channel dimension, power input, temperature of operation, and generated working fluid pressure and fluid flows.

2. EXPERIMENTAL SECTION

This section discusses the fabrication and test procedures for the capillary boiler. Specific microfabrication processes as well as test setups and procedures are reviewed.

2.1. Fabrication

Boilers were constructed of two primary components: a silicon-based boilerplate and an acrylic-based steamdome with volumes for working fluid reservoirs. A silicon substrate formed the basis for each of the boilerplates. Fabrication of the individual boilerplates focused on typical MEMS fabrication processes conducted in a cleanroom environment. Each had an overall size of 13 mm in width by 38 mm in length. This included the primary phase change/heat exchange area as well as area for the working fluid reservoirs at either end as noted in Figure 3. Figure 4 shows a completed boilerplate.

Standard 400 μm thick, double-side polished <100> silicon wafers served as the base for boiler fabrication. Boilers required the manufacture of capillary channels extending between the working fluid reservoirs. Boilers were fabricated with a range of channel depths as part of this work. Previous work established the ability to vary capillary pumping rates as a function of capillary channel height and width dimensions [12,18]. Due to this effect, multiple channel heights were fabricated to fully characterize boiler behavior as a result of capillary pumping variations. Despite changing channel depths, channel widths were maintained at 100 μm to facilitate consistent comparison across boiler designs. Overall channel lengths were 25.5 mm centered on the boilerplate between reservoirs. Figure 4 shows the top-down view of a completed boilerplate and microchannels.

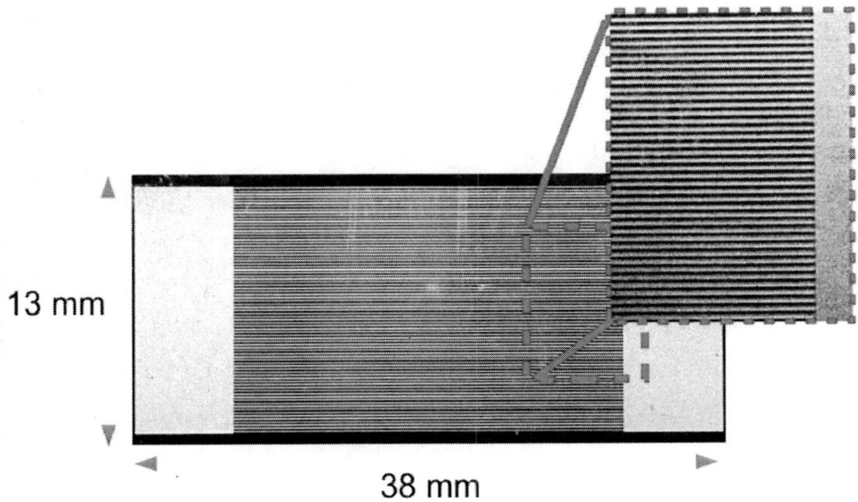

Figure 4. Completed boilerplate.

Channel fabrication was achieved by two different fabrication techniques. For depths below 100 μm, photoresist was spun and patterned atop the silicon wafer in preparation for Reactive Ion Etch (RIE). This produced a photoresist mask in the desired channel pattern atop the silicon wafer. Different channel depths were achieved by varying the RIE exposure time. For fabrication of channels with depths of 100 μm and greater an alternate procedure was used. First, e-beam evaporation was used to deposit a chrome layer atop the silicon wafer to a depth of 200 nm. Photoresist was used to pattern the chrome layer, deposited in a standard spin-coating procedure. Once fully developed, the wafer was placed into chrome etchant to create the final chrome RIE mask utilized to form channels. The chrome proved a more durable mask compared to the photoresist and was able to withstand longer durations of RIE exposure. This allowed the formation of channels with depths greater than 100 μm. Following RIE, any remaining chrome was removed via chrome etchant.

After channels were etched into the silicon wafer, individual boilerplates were diced from the wafer and were ready for incorporation with the acrylic steamdome. Four channel depths were investigated as part of this work. These included depths of 57, 100, 150, and 170 μm.

For each boilerplate manufactured using these processing steps, a top steamdome was added as noted in Figure 3. This acrylic top piece was designed to fit directly over the boilerplate and provide working fluid reservoirs aside the capillary channels. In addition, the central volume of the acrylic piece was left open to serve as a central steam collection point, or steamdome, over the primary boiler channels. 3 mm thick inner walls separated the steamdome from reservoirs with only the boilerplate capillary channels running beneath. Acrylic steamdomes were fabricated using a casting technique where a single prototype was used to form a rubber mold. A two-part liquid resin from Alumilite Corporation was mixed and used to fill the rubber mold. The mixture was allowed to cure for several hours before removal from the mold. This technique produced domes with good repeatability and uniformity. The overall size of the steamdomes (including reservoirs) was 13 mm by 38 mm by 10 mm in height. Small holes were drilled in the top of the acrylic piece. A 1/16" OD hole was drilled atop the reservoir openings to facilitate the addition of working fluid during testing. A larger 1/8" OD hole was drilled atop the main dome outlet and a brass nipple fitting from McMaster-Carr was epoxied in place. This allowed thermocouples and a small pressure sensor to attach to each fully assembled device. 3/16" OD tubing was used for these connections.

Following these additional fabrication steps, each acrylic steamdome was permanently epoxied to a silicon boilerplate. This readied the boiler for testing. Figure 5 shows a steamdome and boilerplate, ready for incorporation of sensors and full testing.

Figure 5. Steamdome and boilerplate.

2.2. Test Setup

Testing of the assembled boilers was conducted based on the boiler's ability to generate working fluid phase change and pressure as a function of input power. The fabricated capillary channels were critical to the operation of the boiler and the generation of a fluid pumping effect. Boilers with varying channel depths were tested and characterized based on these operating parameters.

A common setup was utilized for each boiler tested. In addition, each assembled boiler was tested multiple times to ensure accurate reporting of operating characteristics. The working fluid utilized in these tests was 3M™ HFE-7200. This working fluid was selected for multiple reasons. First, the fluid and channels presented a high degree of wetting, suitable for capillary action. The use of other fluids could vary measured pumping rates and boiler effectiveness depending on fluid-surface contact angle and resulting capillary forces. Other types of fluids would have to be evaluated using this performance metric. Second, the working fluid presents a low boiling point of 76 °C at atmospheric pressure. This is appropriate for many waste heat applications that output thermal energy to the environment at relatively low temperatures. It is important to recognize, however, that the boiler and overall micro-system is not fluid specific. Operation of the system may be tailored to specific thermal sources through a change in working fluid with different boiling point. A final benefit of this working fluid is the small environmental impact potential in any real-world application.

Several components were added to assembled boilers to fully monitor and characterize the performance. Boiler operation involved the use of heat addition via a resistance heater. This heater substituted for potential waste heat sources in a real-world application. The heater was a KHLV-101/P from Omega Engineering. This was hooked to a DC power supply. Both voltage and current consumption were monitored allowing determination of power input to the boiler during testing. A thermocouple (T/C) was added within the boiler steamdome to monitor operating temperature of vapor within the boiler. An additional T/C was located on the heater to measure input temperature. Both thermocouples were k-type, purchased from Omega Engineering. Finally a Precision-Measurements Co (PMC) #060 pressure transducer was incorporated above the steamdome to measure the operating pressure of the boiler. Figure 6 shows the fully assembled boiler and test hardware. The pressure transducer was located as part of the central steamdome exit.

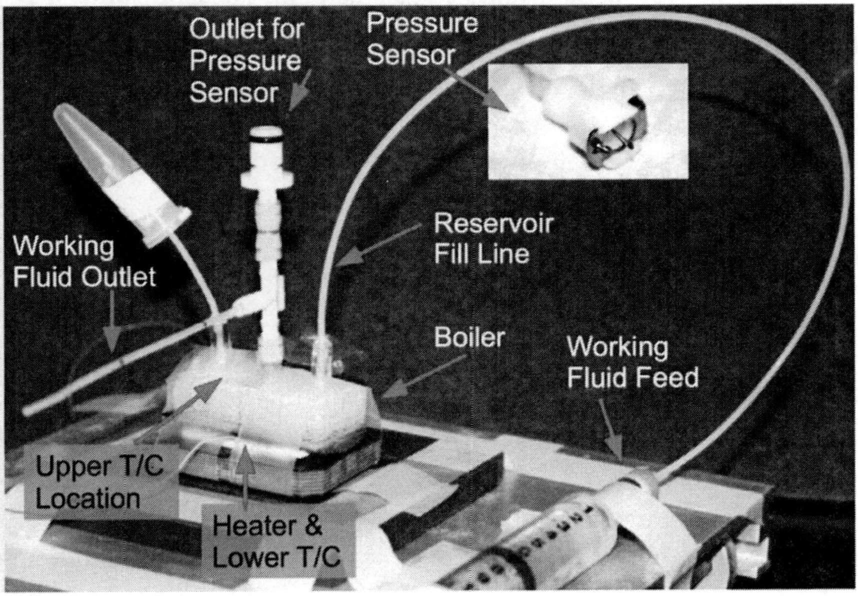

Figure 6. Boiler test setup.

Two basic tests were performed to fully characterize boiler performance. The first allowed vaporized working fluid to freely exit the boiler in an atmospheric condition. This produced results that indicated capillary channel effectiveness directly when assembled with the steamdome. This technique did not produce pressure within the boiler, however. Based on the performance of boilers operating at atmospheric condition, a second test was performed that restricted the exit tube to allow the boiler to operate at a higher pressure. The boiler selected for pressurized operation showed the best performance during atmospheric testing. These second tests produced true boiler-type output consisting of pressurized fluid flow. The direct operating procedure for both tests was similar.

In each test, thermal energy was applied to the boiler for several minutes without working fluid present. This thermal energy simulated heat transfer from a real-world source. This allowed the boiler to achieve a steady state condition prior to working fluid addition. Power inputs considered for these tests were 2.6 W and 3.6 W as measured at the resistance heater. This maintained boiler operation within or below an envelope of thermal input consistent with potential real-world sources. Tests conducted at elevated pressures considered only the elevated input of 3.6 W. Steady state operation was determined by monitoring thermocouple readout. After achieving steady state, the boiler was loaded with working fluid via syringe connected to the working fluid reservoirs as noted in Figure 6. This provided the needed working fluid to the capillary channels for phase change. The feed was connected via sealed connection to prevent any pressurized working fluid generated by the boiler from backing into the working fluid syringe feed. In this manner, vapor was forced to exit the centrally located steamdome hole.

Fluid was supplied in two ways depending on the type of test being conducted. For testing conducted in atmospheric conditions, experiments were conducted that illuminated the basic capillary flow rate capabilities of the boilers. To achieve this, reservoirs surrounding the boilerplate were slightly oversupplied to allow the capillary channels a maximum amount of fluid for phase change. By contrast, tests conducted under pressurized conditions used these achieved capillary flows as a guideline and did not oversupply working fluid. This had the desirable effect of focusing boiling and phase change within the channels during pressurized tests. This also eliminated potential pool boiling resulting from over-full conditions. In both cases, a syringe pump was sourced to predictably compress the supply syringe. This maintained working fluid within the boiler at controllable and predictable levels.

Further advantages of syringe pump use included elimination of large-scale temperature fluctuations caused by unpredictable supply. In addition, operation in this manner represented an improved approximation of boiler operation as part of a larger system. In a full system, working fluid would be returned to the boiler at predictable rates based on the boiler output flow rates and consumption by any attached power output device.

Independent of test type, following the addition of working fluid the boiler was allowed to operate at a new steady state operating condition. As with the initial dry "warmup," this new steady state condition was determined based on a consistent operating temperature of the boiler. Figure 7 shows a typical temperature plot from within the steamdome following the addition of working fluid. The constant temperature section represented the period considered for testing and analysis presented in this paper. Individual tests were conducted using boilerplates with capillary channels of 57, 100, 150, and 170 μm depth. Outcome from these tests represented a significant characterization of boiler operation across multiple power inputs using multiple designs.

Independent of test type, vaporized working fluid was allowed to exit the boiler via the central steamdome hole in line with the pressure transducer. The working fluid was directed into a short length of tubing. For atmospheric testing, focused primarily on capillary channel effectiveness within the steamdome, the condensed working fluid was collected within a syringe. The total volume was measured as a function of test duration yielding specific flow rate information. This mass flow rate was combined with the known heat of vaporization for HFE 7200, 125 kJ/kg at atmospheric pressure. This allowed determination of power absorption via phase change of the working fluid.

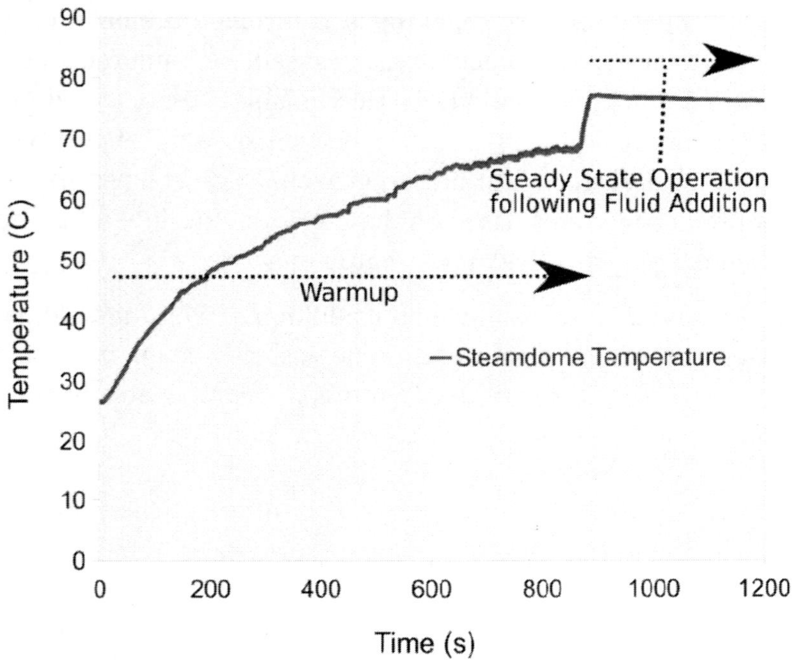

Figure 7. Typical temperature plot and test section.

Following tests within the atmospheric pressure steamdome, pressurized testing was conducted. The boiler was brought to a dry, steady state condition as before and the syringe pump was allowed to input working fluid to the reservoirs as previously described. A restriction in the form of a clamping force was placed on the central exit tube to increase the boiler operating pressure. This effectively simulated operation when the boiler must generate operating pressure for attached devices. The boiler was operated for an additional several minutes and allowed to come to a steady state under these new conditions. Based on previously established guidelines, the feed rate of the syringe pump was adjusted as needed such that working fluid levels within the boiler were maintained at a constant value. Given a change in operating pressure, the feed rate was adjusted accordingly to ensure no oversupply and pool boiling.

This was confirmed both visually and via pressure and thermocouple readout. In this manner, the boiler was operated at a steady, pressurized condition with a known mass flow rate from the working fluid supply. By varying the clamping force, different operating pressures were achieved and capillary action was forced to work against generated boiler pressures. Power input was maintained at 3.6 W for pressurized tests.

Independent of test type, T/C and pressure sensor readouts were recorded via Labview software. This raw data was post-processed to formulate the final test results.

3. RESULTS AND DISCUSSION

Tests were conducted as described in Section 2. These tests resulted in determination of individual boiler operation based on both heat input and boilerplate channel design. In the first set of tests, the boiler was allowed to operate at an atmospheric pressure condition where no restriction was placed on the main exit tube. In the second test, a restriction was in place and the boiler was operated at elevated pressures. First, results are reported for the atmospheric system given 2.6 W input. Second, atmospheric results from 3.6 W input are reported. Finally, operation of the boiler at elevated pressure is reported. Key measurements recorded during testing were the working fluid mass flow rate, the operating pressure of the boiler, temperature of the simulated thermal source as provided by the resistance heater, and internal steamdome operating temperature of the vapor. Table 1 and Table 2 summarize atmospheric operation of the boiler. Mass flow measurement error was determined to be on the order of ±2.5μg/s given the accuracy of measurements employed.

Table 1. Atmospheric boiler operation at 2.6 W thermal input.

Channel Depth (μm)	Source Temperature (°C)	Steamdome Vapor Temperature (°C)	Mass Flow (mg/s)
57	94.2	74.3	4.53
100	106.9	74.8	4.56
150	101.5	75.8	4.61
170	103.4	76.2	4.61

Table 2. Atmospheric boiler operation at 3.6 W thermal input.

Channel Depth (μm)	Source Temperature (°C)	Steamdome Vapor Temperature (°C)	Mass Flow (mg/s)
57	105.3	74.6	8.86
100	121.6	75.9	10.66
150	116.2	76.1	11.47
170	120.5	76.1	12.26

Several observations may be made based on these initial runs. First, as expected the varying channel depths exhibit different mass flow rates within the steamdome. This was consistent with other devices and base experimental and modeling activities conducted by Kim *et al.* [11,16]. A general trend was established where boilers constructed with large channel depths showed increasing internal fluid flow resulting in increased rates of phase change. Given 2.6 W input, the mass flow was increased by 1.7% when channels increased from 57 μm depth to 170 μm depth. These effects were amplified at increased thermal inputs. Given 3.6 W input, mass flow increased by nearly 28% as channel depth increased from 57 to 170 μm. Figure 8 shows the recorded mass flow rates *versus* channel depth for the boilers tested.

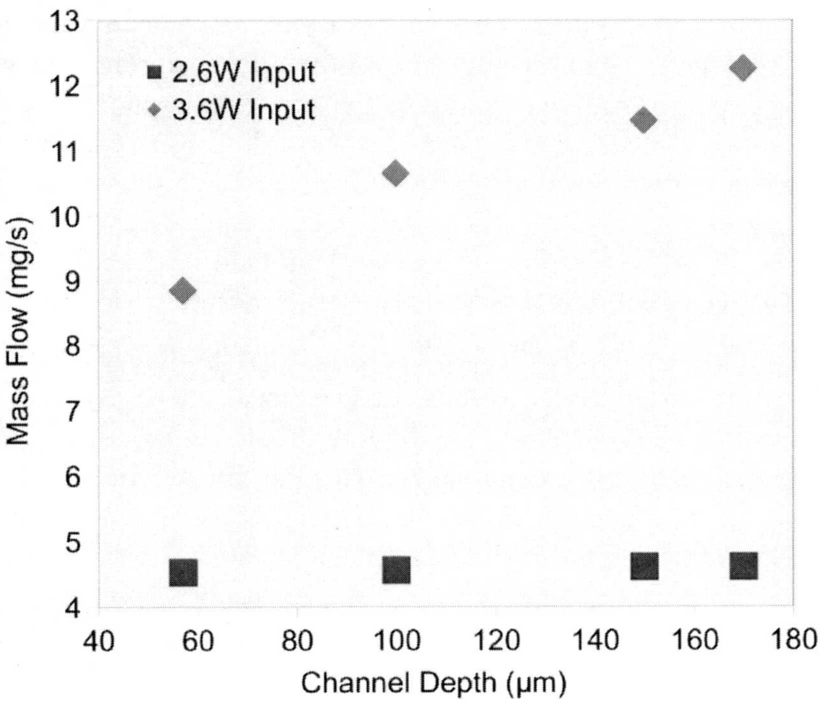

Figure 8. Mass flow *vs.* channel depth.

In addition to varying mass flow rates, it is worth noting the operating temperatures inside the steamdome. The T/C was positioned such that it was truly suspended within the vapor contents of the steamdome. Readings reflected the boiling point of the working fluid at atmospheric pressure, about 76 °C. This confirmed the saturated environment within the boiler.

Energy absorption via phase change was determined based on these results. Given the working fluid heat of vaporization, 0.57 mW was consumed by the 57 μm boiler given 2.6 W input. Energy absorption was 0.58 mW for 170 μm channels at the same power input. Amplified trends were observed given 3.6 W thermal input. Power consumption via phase change was 1.11 mW for the 57 μm channel boiler. 1.53 mW was consumed by the 170 μm channel boiler.

The power consumed via phase change yielded poor conversion efficiencies given the heat transferred to the boiler during testing. This is explained through consideration of the area and thermal mass of the enlarged steamdomes. Although constructed from acrylic-type material, the domes represent a heat sink, allowing fluid to condense before exiting the system via centralized output hole. While an effective means to collect working fluid from the boiler, future designs will work to limit the dome effect on premature condensation within the chamber itself.

This is evidenced by the results. At increased power input, the average temperature of the steamdome walls should be higher due to elevated boilerplate temperatures. This limited the effect of the steamdome as a thermal heat sink. As a result, the boiler showed greatly increased mass transfer independent of channel dimension at these elevated power inputs. In addition, the specific advantages of increasing channel depths were greatly amplified across the range of channels considered.

Following investigations at atmospheric pressure, the outlet from the boiler was restricted. This had the effect of increasing the operating pressure of the boiler for a given heat input. Several combinations of pressure and flow rate were examined. Operating pressures were increased in several steps using ambient tests as a baseline (matching atmospheric testing in Table 2). As discussed in Section 2.2, the boiler with 170 μm channel depths was selected due to its preferred performance in atmospheric testing. Table 3 shows the operating pressures with associated flow rates and steamdome temperatures recorded during pressurized testing.

Table 3. Boiler operation at increased pressure and 3.6 W input.

Channel Depth (μm)	Steamdome Vapor Temperature (°C)	Mass Flow (mg/s)	Pressure (Pa)
170	75.7	12.3	0 g
170	76.3	10.8	19.3
170	76.7	8.6	55.0
170	78.3	5.4	229.3

Results of these tests revealed a maximum operating pressure of 229 Pa given the fluid supply rates. These rates were guided by atmospheric tests and adjusted to ensure pool boiling and over-supply was avoided. Flows were routinely reduced to match the increasing pressure load of the boiler. This makes sense due to the inherent restriction placed on the central steamdome exit hole. The temperature of the vapor within the steamdome was also observed to increase when the boiler was operated at elevated pressure. This correlates to an increasing saturation temperature corresponding to the increased surrounding steamdome pressure. Figure 9 explores this trend directly and illustrates the boiler operating with varying pressure settings. The maximum pressure achieved for this test was 275 Pa, although was not considered as a steady state point for true performance reporting. As pressure varied, the temperature followed these specific pressure fluctuations demonstrating the saturated operating environment achieved within the boiler.

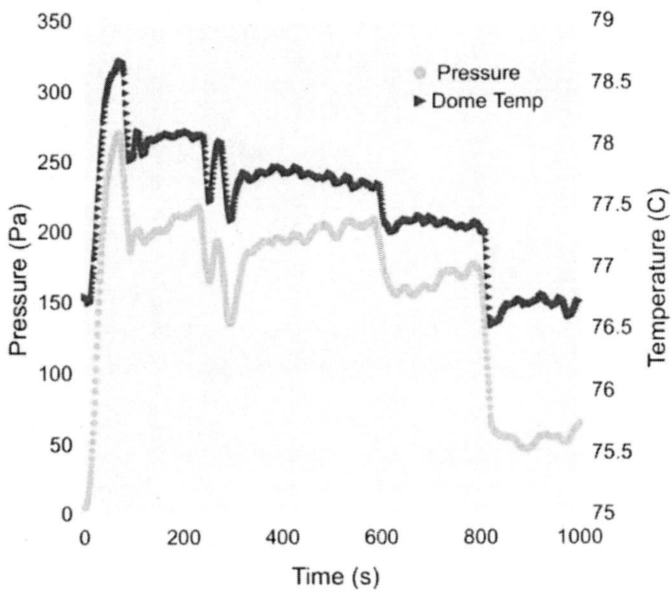

Figure 9. Temperature and steamdome vapor pressure in boiler.

Overall boiler outputs were governed by the same constraints as presented by the steamdome in atmospheric testing. As a result, achieved pressures and flow rates were hampered by the steamdome heat sink. It is fully expected future steamdome redesign efforts will enhance both pressure and mass flow rate capabilities of these devices.

It is useful to put these tests and thermal inputs into a real world context. Given the boilerplate area, the input amounts resulted in power densities of 5.0 kW/m^2 and 7.8 kW/m^2. These values prove to be conservative estimates compared to potential sources like a typical Otto cycle device. A typical 4 cylinder Otto cycle engine has an expected thermal energy loss directly from the engine sides on the order of 10% supplied fuel energy [19]. Given an engine output rating of 150 kW, this equates to 45 kW loss directly to engine surroundings. Using typical dimensions for an engine of this type, this is on the order of 30 kW/m^2. Based on this example, there is clear potential to drive a boiler like those presented in this work. By varying the working fluid, boiler operation may be tuned to specific energy inputs and temperature profiles.

4. CONCLUSIONS

For the first time, a novel capillary-action based boiler has been fabricated and tested. The design of the boiler is unique, dependent on capillary action to drive fluid flow across the heated boilerplate surface. In this device, heat may be continuously transferred to the boiler from an external source like waste heat or available environmental thermal sources. As part of a complete microfabricated system, useful power may be generated.

Boilerplates were fabricated from silicon substrates with capillary channels of 100 μm width. Channel depths were varied as part of these investigations using values of 57, 100, 150, and 170 μm. Fluid flow rates, operating pressures, and internal working temperatures were monitored during these tests. In atmospheric pressure testing, the best mass transfer recorded was via 170 μm channel and 3.6 W thermal input. The mass transfer rate was 12.26 mg/s under these conditions. This was an improvement of 28% compared to boilers constructed with 57 μm depth capillary channels. The total power consumed via phase change in these devices was 1.53 mW. Pressurized testing was conducted at pressures ranging from atmospheric to 229 Pa. Boilers with 170 μm depth channels were selected for these tests based on their performance in atmospheric conditions. Mass flow rates were varied according to operating pressure, ranging from 12.3 mg/s at atmospheric pressure to 5.4 mg/s at 229 Pa.

Several opportunities for improved operation were noted as a result of these tests. The use of the steamdome was an effective method for pressure generation and working fluid collection but resulted in significant cold surface for condensation within the boiler itself. The use of smaller volumes will be investigated in future boiler constructions. There is also significant opportunity to increase the base thermal conductivity of the boilerplate via alternate material investigations and channel constructions. The use of metallic channels, for example, has the potential to increase the net thermal transfer to the working fluid during device operation. These modifications should produce a more efficient system, capable of operation from low-temperature waste heat sources. Despite the challenges noted, this work illustrates the promise of this unique approach to low-exergy energy use using a small-scale microfabricated system.

ACKNOWLEDGMENTS

The authors gratefully acknowledge the support of this work by the NSF via Grant No. ECCS-1053729, and the State of Louisiana Board of Regents.

REFERENCES

1. Energy Information Administration (EIA). *Annual Energy Review 2011*, Available online: http://www.eia.doe.gov/emeu/aer/contents.html (accessed on 16 March 2012). Technical Report 2011.

2. *2. Annual Energy Outlook 2009 with Projections to 2030; Technical Report DOE/EIA-0383*; Department of Energy, National Energy Information Center, EI-30 Energy Information Administration Forrestal Building: Washington, DC, USA, March 2009.

3. Heywood, J.B. *Internal Combustion Engine Fundamentals*; McGraw-Hill Inc.: New York, NY, USA, 1988; pp. 733–736.

4. Ogbonnaya, E.; Gunasekaran, A.; Weiss, L. Fabrication of a MEMS-based solar thermal collector for energy sustainability. In Proceedings of the 2011 ASME IMECE, Denver, CO, USA, 11–17 November 2011; pp. 1–6, No IMECE2011-63842.

5. Lipkin, R. Micro steam engine makes forceful debut. *Sci. News* **1993**, *144*, 197.

6. Lee, C.; Liamini, M.; Frechette, L.G. A silicon microturbopump for a rankine-cycle power-generation microsystem—Part II: Fabrication and characterization. *J. Microelectromech. Syst.* **2000**, *20*, 326–338.

7. Lee, C.; Fréchette, L.G. A silicon microturbopump for a rankine-cycle power generation microsystem—Part I: Component and system design. *J. Microelectromech. Syst.* **2011**, *20*, 312–325.

8. Muller, N.; Frechette, L. Performance analysis of Brayton and Rankine cycle microsystems for portable power generation.

In Proceedings of the 2002 ASME IMECE, New Orleans, LA, USA, 17–22 November 2002; pp. 513–522.

9. Epstein, A.H. Millimeter-scale, micro-electro-mechanical systems gas turbine engines. *J. Eng. Gas Turbines Power* **2004**, *126*, 205–226.

10. Lee, C.K.; Jiang, P.; Prewett, P. Design and fabrication of a micro wankel engine using MEMS technology. *Microelectron. Eng.* **2004**, *73–74*, 529–534.

11. Epstein, A.H.; Jacobson, S.A.; Protz, J.M.; Frechette, L.G. Shirtbutton-sized gas turbines: The engineering challenges of micro high speed rotating machinery. In Proceedings of the 8th Int'l Symposium on Transport Phenomena and Dynamics of Rotating Machinery (ISROMAC '08), Honolulu, HI, USA, January 2000.

12. Weiss, L.; Richards, C.; Richards, R. Power output force generation by a MEMS phase change actuator. *J. Microelectromech. Syst.* **2011**, *20*, 1287–1297.

13. Weiss, L. Resonant operation and cycle work from a MEMS-based micro-heat engine. *Microsyst. Technol.* **2009**, *15*, 485–492.

14. Bardaweel, H.K.; Anderson, M.J.; Weiss, L.W.; Richards, R.F.; Richards, C.D. Characterization and modeling of the dynamic behavior of a liquid-vapor phase change actuator. *Sens. Actuat. A: Phys.* **2009**, *149*, 284–291.

15. Cho, J.; Wiser, T.; Richards, C.; Bahr, D.; Richards, R. Fabrication and characterization of a thermal switch. *Sens. Actuat. A: Phys.* **2007**, *133*, 55–63.

16. Cho, J.; Weiss, L.; Bahr, D.; Richards, C.; Richards, R. Power production by a dynamic micro heat engine with an integrated thermal switch. *J. Micromech. Microeng.* **2007**, *17*, S217–S223.

17. Weiss, L.W. Study of a MEMS-based free piston expander for energy sustainability. *J. Mech. Des.* **2011**, *132*, 091002–091008.

18. Kim, D.; Lee, K.; Kwon, T.; Lee, S. Micro-channel filling flow considering surface tension effect. *J. Micromech. Microeng.* **2002**, *12*, 236–246.

19. Ferguson, C.; Kirkpatrick, A. *Internal Combustion Engines*, 2nd ed; John Wiley and Sons Inc.: New York, NY, USA, 2001.

5

CHAPTER

INFLUENCE OF ORIENTATION AND RADIATIVE HEAT TRANSFER ON ALUMINUM FOAMS IN BUOYANCY-INDUCED CONVECTION

Marijn Billiet *, Sven De Schampheleire, Henk Huisseune and Michel De Paepe

Department of Flow, Heat and Combustion Mechanics, Ghent University, Sint-Pietersnieuwstraat 41, Ghent 9000, Belgium

ABSTRACT

Two differently-produced open-cell aluminum foams were compared to a commercially available finned heat sink. Further, an aluminum plate and block were tested as a reference. All heat sinks have the same base plate dimensions of four by six inches. The first foam was made by investment casting of a polyurethane preform and has a porosity of 0.946 and a pore density of 10 pores per linear inch. The second foam is manufactured by casting over a solvable core and has a porosity of 0.85 and a pore density of 2.5 pores per linear inch. The effects of orientation and radiative heat transfer are experimentally investigated. The heat sinks are tested in a vertical and horizontal orientation. The effect of radiative heat transfer is investigated by comparing a painted/anodized heat sink with an untreated one. The heat flux through the heat sink for a certain temperature difference between the environment and the heat sink's base plate is used as the performance indicator. For temperature differences larger than 30° C, the finned heat sink outperforms the in-house-made aluminum foam heat sink on average by 17%. Furthermore, the in-house-made aluminum foam dissipates on average 12% less heat than the other aluminum foam for a temperature difference larger than 40° C. By painting/anodizing the heat sinks, the heat transfer rate increased on average by 10% to 50%. Finally, the thermal performance of the horizontal in-house-made aluminum foam heat sink is up to 18% larger than the one of the vertical aluminum foam heat sink.

Keywords: experimental; heat sink; natural convection; metal foam; finned heat sink; radiation; heat transfer

1. INTRODUCTION

Electronic components are omnipresent in daily life products, such as cars, computers and power converters. Hence, it is important to reduce the chance of premature electronics failure. The main causes of electronics failure are over-voltage, elevated operating temperature, moisture and electrostatic discharge (ESD) [1]. To avoid failures caused by elevated operating temperatures, electronic components are cooled. Besides the increased lifetime, some components, such as LEDs, are also more efficient at lower operating temperatures [2,3,4].

Due to the miniaturization of electronic devices and the resulting rise of heat fluxes [5,6], electronic devices need to be cooled by heat sinks. Two types of heat sinks can be distinguished: passive and active heat sinks. The passive heat sink dissipates heat by buoyancy-induced convection. The active heat sink is equipped with a fan to induce an air flow through it and thereby dissipates heat by forced convection. From an economic and ecological point of view, passive heat sinks are favorable over active heat sinks, because they are less expensive, more reliable, require less maintenance and do not use energy.

Current heat sinks use fins to increase their surface-to-volume ratio. These fin shapes can become quite complex, e.g., three-dimensional surface extensions [7,8,9]. One promising material that can be used as three-dimensional surface extensions is high-porous open-cell metal foam.

Open-cell metal foam is a porous media that has a very high volumetric porosity ϕ which is defined as:

$$\phi = \frac{V_{\text{total}} - V_{\text{solid}}}{V_{\text{total}}} \quad (1)$$

The volumetric porosity is typically larger than 0.85. A high porosity will be beneficial for mobile applications due to the reduced weight of the heat sink.

The structure of high-porous open-cell metal foam can be found in Figure 1. The metal foam consists of cells that are connected through pores. Each cell is made of several struts, which connect in the nodes. The cell size of metal foam is commonly given as the pores per linear inch (PPI)-value. The porosity ϕ and the PPI-value are typically used to characterize open-cell metal foam in the literature. Nevertheless, other methods for characterizing metal foam exist [10].

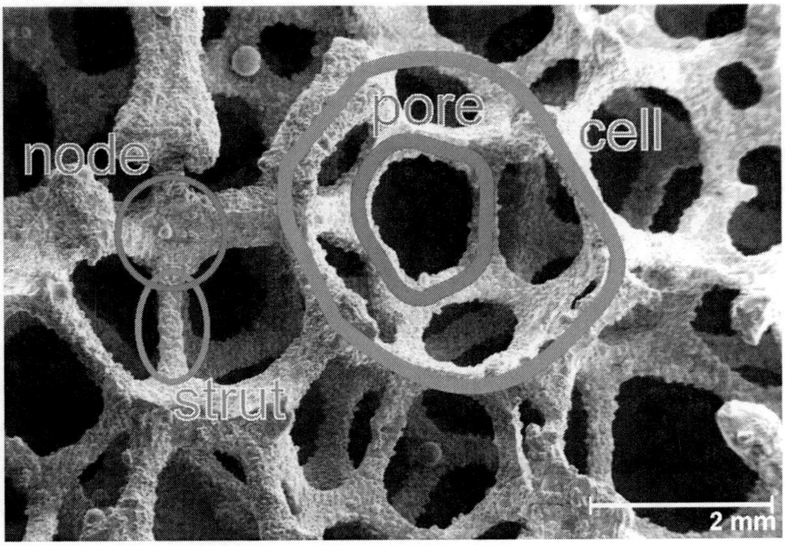

Figure 1. The structure of open-cell metal foam.

The work of Ashby [11], Banhart [12] and Dukhan [13] describes the different production methods of open-cell metal foam. Two methods that are relevant to this work are investment casting of a polymer foam precursor and leachable bed casting. A product example of each production method is shown in Figure 2. In the first method, an open-cell polymer foam is made by reticulation of a polyurethane foam [14]. Then, the investment casting technique is used to make an exact metallic copy of this open-cell polymer foam. In the second method, metal is cast over a stacked bed of solvable spheres, e.g., salt spheres or sand with a polymer bonding agent [15]. After solidification, these spheres are removed.

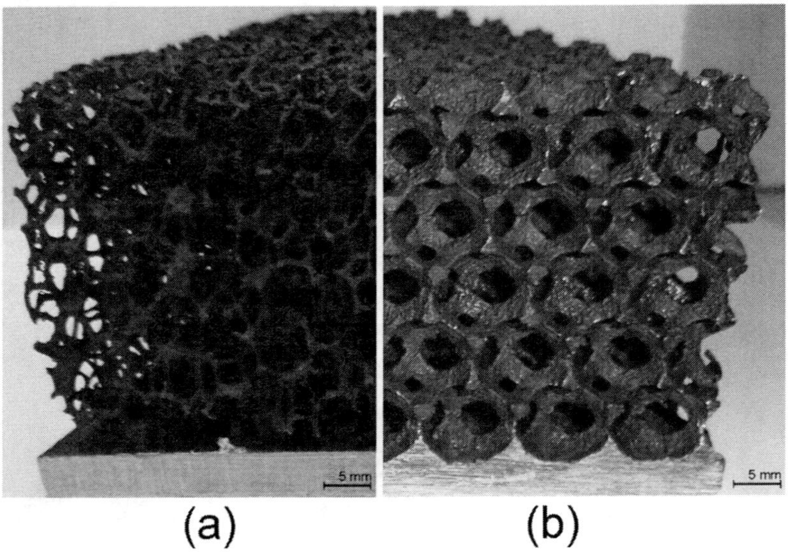

Figure 2. An example of a metal foam produced by (a) investment casting of a polymer foam precursor and (b) leachable bed casting.

To date, metal foam used in forced convection is extensively described in the literature [16,17,18]. However, open literature considering buoyancy-induced convection is limited.

Aluminum foam in buoyancy-induced convection was first investigated by Bhattacharya et al. [19]. The authors tested different aluminum foam heat sinks in a horizontal and vertical orientation. Furthermore, their finned aluminum foam heat sink was compared to a commercially available finned heat sink. Qu et al. [20] developed a heat transfer correlation for copper foam in buoyancy-induced convection at an arbitrary orientation. Besides the orientation, the correlation also included the effects of the foam structure and the height of the heat sink. Recently, De Schampheleire et al. [21,22] investigated the influences of porosity, pore size, foam height and radiative heat transfer on the thermal performance of aluminum foam heat sinks in buoyancy-induced convection, both numerically and experimentally. Based on these results published in the open literature, the following conclusions can be drawn:

- The heat transfer of highly-porous metal foam in buoyancy-induced convection increases with decreasing porosity [19,20,21];
- The heat transfer of highly porous metal foam in buoyancy-induced convection decreases with increasing pore density [19,20,21];
- The heat dissipated by the metal foam heat sink increases with increasing heat sink height until it reaches an asymptotic value. The asymptotic value is reached when the effect of the increasing viscous drag and the temperature drop along the heat sink's height equals the effect of increasing surface area [20,21];
- De Schampheleire et al. [22] indicated the importance of the radiative heat transfer of metal foam heat sinks in buoyancy-driven convection. The authors found that up to 30% of the total heat transfer of aluminum foam heat sinks in buoyancy-induced convection is due to thermal radiation;
- The experimental data of both Bhattacharya et al. [19] and Qu et al. [20] suggest a non-significant effect of the orientation on the thermal performance of metal foam heat sinks.

The first aim of this study is to compare the thermal performance of an aluminum foam heat sink made by investment casting of a polymer precursor, an aluminum foam heat sink made by leachable bed casting and a finned heat sink in buoyancy-induced convection. To the best of our knowledge, an experimental comparison between a finned heat sink and an aluminum foam heat sink has not yet been done. Bhattacharya et al. [19] only compared a finned metal foam heat sink with the correlation of an optimal finned heat sink and concluded that the finned metal foam heat sink performed better. Further, the influence of thermal radiation on buoyancy-induced convection is investigated. In contrast to the work that has already been done, the influence of the foam production method and the height of the foam was investigated. Finally, this paper investigated the influence of the heat sink's orientation on its thermal performance.

2. METHODOLOGY

2.1. Experimental Setup

No standardized methodology exits for testing heat sinks in buoyancy-induced convection. Hence, the design of the experimental setup was based on experimental setups found in the literature [19,20,21,23,24]. The design of the experimental setup aimed to achieve a one-dimensional heat flux through the heat sink.

The experimental setup, depicted in Figure 3, consists of a frame, a rotation mechanism and an insulated box. The rotation mechanism enables the setup to test the heat sinks in different orientations.

Figure 4 and Figure 5 are a top view and a cross-section of the insulated box, respectively. To test different heat sink sizes in future work, four heater elements with different dimensions are placed in an array. The heat sink that will be tested is placed on top of one or multiple heater elements. In the current study, only the 4″ × 6″ heater element was used.

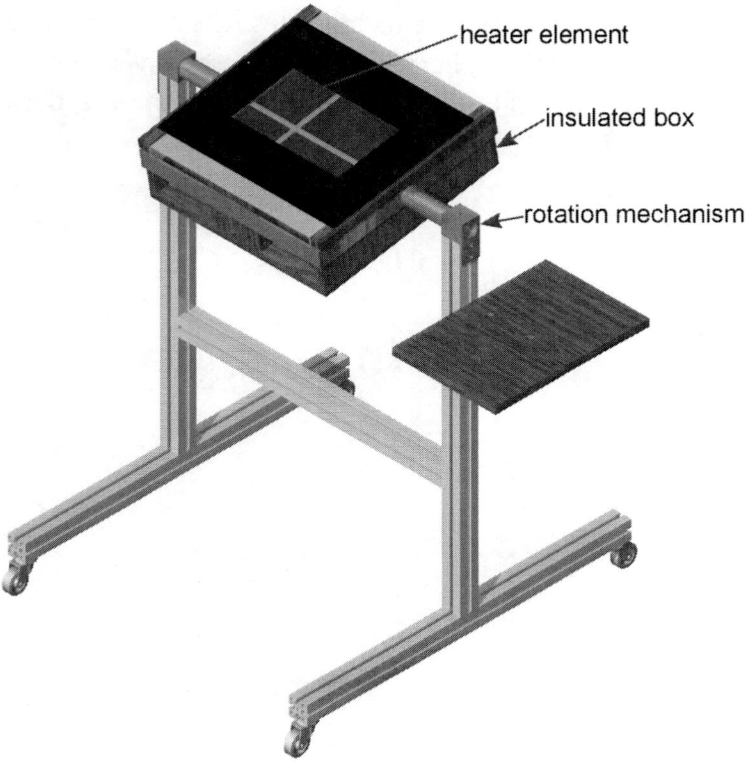

Figure 3. Overview of the experimental setup.

Figure 6 represents the cross-section of one of the four heater elements. This heater element consists of a stack of metal plates and an electrical film heater (Omega® KH series, OMEGA Engineering, Inc., Stamford, CT, USA). The two copper plates ensure a uniform temperature distribution at the base plate of the heat sink, resulting in a uniform heat flux through the base plate. The uniform temperature distribution was verified using an IR-camera. The temperature uniformity was within the uncertainty of the camera. The copper plates are also used to fix the thermocouples in between them. The aluminum plate is added to compensate for different base plate thicknesses. Further, layers of thermal paste (k = 0.8 W/(m.K)) and thermal pads (k = 6 W/(m.K)) are added in between to avoid insulating air gaps between the metal plates. The two largest and the two smallest electrical film heaters are each respectively powered by an Elektro Automatik® PS 8160-04 2U (Helmholtzstrasse, Germany) and a TTi® PLH120-P (Fort Worth, TX, USA).

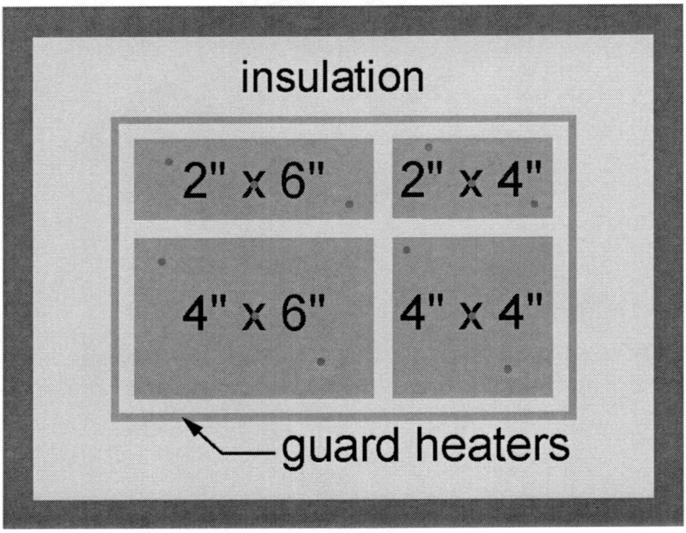

Figure 4. Top view of the insulated box. The blue dots indicate the locations of the thermocouples.

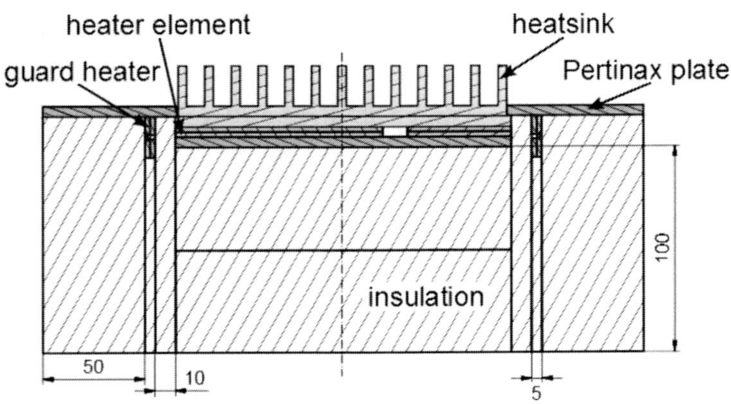

Figure 5. A cross-section of the insulated box.

As depicted in Figure 5, guard heaters are only used at the sides, while the bottom surfaces of the heater elements are well insulated. The insulation material used is Microtherm (k = 0.0221 W/(m.K)). Below the main heaters, there is 100 mm of insulation. The thickness of insulation next to the guard heaters is 50 mm and in between the guard and main heater there is 10 mm of insulation.

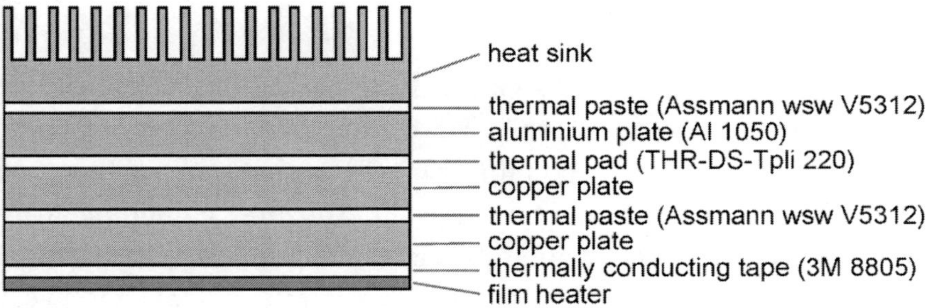

Figure 6. Cross-section of a heater element.

The side guard heaters consist of 2 copper plates and an electrical film heater (Omega® KHLV series). These side guard heaters are powered by a TTi® PL303QMD-P. When only part of the main heaters is used, which is the case in this study, the other heater elements serve as extra guard heaters.

The heat sink is firmly pressed against the heaters by a 5 mm-thick plate of Pertinax® (k = 0.2 W/(m.K)), as shown in Figure 5. In addition, this plate forms one smooth surface with the base plate of the heat sink to avoid any disturbance of the air flow.

The temperature of each heater element is measured by 3 K-type thermocouples (D j = 0.5 mm) and the temperature of each side guard heater is measured by 2 K-type thermocouples (D j = 0.5 mm). In both cases, the thermocouples are placed between the two copper plates (Figure 5 and Figure 6) into a machined slot of 0.5 mm, which is not visible on the figure. The ambient temperature is measured by 4 K-type thermocouples (D j = 0.75 mm). These 4 thermocouples are placed at the four corners of the experimental setup at a height of 15 cm above the Pertinax® plate (Fibox, Espoo, Finland). All thermocouples are read out by a data acquisition system (Keithley® 2700, Keithley Instruments Inc., Cleveland, OH, USA).

2.2. Test Samples

Table 1 lists the tested heat sinks. All tested heat sinks have the same base plate dimensions (4″ × 6″). First, a flat aluminum plate (= bare base plate) and two solid aluminum blocks were tested. Further, two different aluminum foams were tested. The first aluminum foam (reticulated22 and reticulated40) was made in-house by investment casting of a polymer foam precursor. This aluminum foam has a very high porosity (94.6%) and a moderate PPI-value (10 PPI). Both values are given by the manufacturer. The average cell diameters measured in two perpendicular directions are respectively 3.4 mm and 4.8 mm. The measured pore density is 6 PPI in one direction and 8 PPI in the perpendicular direction. This value is lower than specified by the manufacturer. The second aluminum foam tested (spheres40) was bought from Alveotec® (Venissieux, France) [25]. This foam was produced using leachable bed casting. This foam has a relatively low porosity (85%) and PPI-value (2.5 PPI). Both values are given by the manufacturer. The average cell diameters measured in two perpendicular directions are respectively 8.1 mm and 8.3 mm. The measured pore density is 2.5 PPI in one direction and 2.6 PPI in the perpendicular direction. Although the Alveotec heat sink consists only of 4 pores in the height direction, the heat sink is representative for the Alveotec foam properties. Brun [26] demonstrated that 4 pores in every direction are enough to construct a representative elementary volume for thermo-hydraulic calculations. For constructing the aluminum foam heat sinks, the aluminum foam was glued onto the base plate using thermal conductive epoxy (Loctite®

ESP110, Düsseldorf, Germany). Finally, a finned heat sink (Type S120), which was produced by Aavid® (Aavid Thermalloy, Laconia, NH, USA) [27], was also put to the test. After the first experiments, all heat sinks, except the finned heat sink, were painted with Kontakt Chemie® (Jena, Germany) Graphit 33. For the finned heat sink, an anodized version was bought and tested.

Table 1. Overview of the test samples.

Name	Width (mm)	Length (mm)	Height (mm)	# Fins	Porosity (-)	PPI	σ (m^2/m^3)	Material
plate	150.3 ± 0.05	101.4 ± 0.05	0	-	-	-	-	Al 1050
block22	150.0 ± 0.05	99.0 ± 0.05	22.0 ± 0.05	-	0	-	79	Al 1050
block40	150.0 ± 0.05	99.0 ± 0.05	40.0 ± 0.05	-	0	-	59	Al 1050
reticulated22	150.0 ± 0.05	100.9 ± 0.05	22.2 ± 0.05	-	0.946	10	486	Al 1050
reticulated40	149.3 ± 0.05	101.0 ± 0.05	40.0 ± 0.05	-	0.946	10	486	Al 1050
spheres40 [25]	149.8 ± 0.05	97.5 ± 0.05	40.0 ± 0.05	-	0.85	2.5	360	Al 1050
finned22 [27]	150.0 ± 0.05	101.0 ± 0.05	22.0 ± 0.05	25	0.65	-	352	Al 6060

2.3. Measuring Procedure

First, the temperature difference between the heat sink's base plate and the ambient air was set between 10° C and 70° C. Then, the side guard heaters' temperature was controlled to the average temperature of the heat sink's base plate using a PID-controller. Finally, when the steady state was reached, the power dissipated by the heat sink was measured as a function of the temperature difference between the base plate of the heat sink and the ambient air. Steady state was attained when the standard deviation of 100 temperature measurements of the heat sink's base plate at a sampling frequency of 0.17 Hz was below 0.02 K. Some of the experiments were repeated to confirm the consistency of the experimental results.

The results are reported as the heat dissipated by the heat sink as a function of the temperature difference between the heat sink's base plate and the ambient air. This is used as a measure for the thermal performance of the heat sink. For the Alveotec® foam heat sink, the heat dissipated by the heat sink per base plate surface area was also calculated to compensate for the small differences in base plate dimensions.

2.4. Uncertainty Analysis

The uncertainty analysis used in this work follows the textbook of Taylor [28].

The dimensions of all components are measured using a caliper with an accuracy of 0.05 mm.

All thermocouples are calibrated using a dry-block calibrator (Druck® DBC 150, Rho, Italy) and a reference thermometer (Fluke® 1523, Everett, WA, USA). After calibration, the average uncertainty of the thermocouples is ±0.11 ∘ C., and the maximum uncertainty is ±0.2∘ C.

$$Q_{\text{hs}} = P_e - Q_{\text{loss}} \tag{2}$$

Equation (2) gives the power dissipated by the heat sink. The power supplied by the electrical heaters (P_e) is measured internally by the power supplies. The relative uncertainty of the power measurement depends on the actual power, which is measured and ranges between 4.7% (at 5 W) and 1.5% (at 50 W).

The heat loss through the insulation (Q_{loss}) is estimated using a 3-dimensional finite elements conduction model. The geometry used in the model resembles the actual experimental setup, shown in Figure 5. Two parts were simplified in the model's geometry. First, the heater elements were replaced by a solid block of copper at a constant temperature. Further, the tested heat sink was modeled as a flat plate with an equivalent convection coefficient. At the other boundaries, a fixed convection coefficient of 5 W/K was assumed. The free air temperature was set to 20 ∘ C, which was comparable to the real testing conditions. To verify the estimation of the heat loss, the results of the aluminum plate experiments were compared to correlations from the literature [29,30] (see Figure 7).

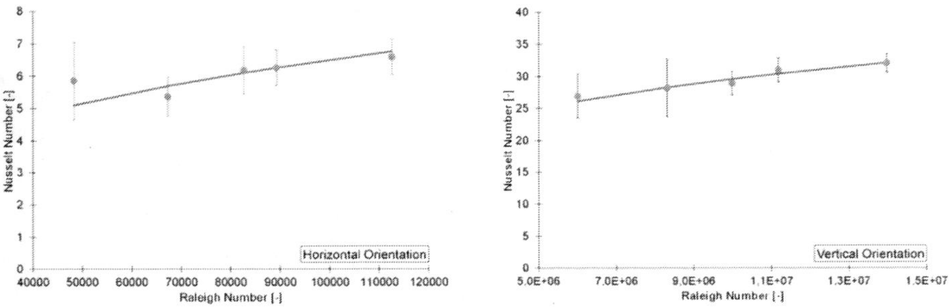

Figure 7. A comparison between a painted flat plate in horizontal and vertical orientation with the correlation found in the literature [29,30].

3. RESULTS AND DISCUSSION

First, the thermal performance of the in-house-made aluminum foam heat sink was compared to that of the finned heat sink and of the solid aluminum block. Later, a comparison between the in-house-made aluminum foam and the Alveotec® foam was made. Then, the influence of painting was investigated. Finally, the effect of orientation on the thermal performance was studied.

3.1. Comparison to Commercial Finned Heat Sink

The in-house-made aluminum foam heat sink (reticulated22) was compared to the finned heat sink (Aavid® S120) and the solid aluminum block in buoyancy-induced convection. All three tested heat sinks have the same height of 22 mm. Figure 8 shows the heat dissipated by the heat sink as a function of the temperature difference between the heat sink's base plate and the ambient air for all three heat sinks.

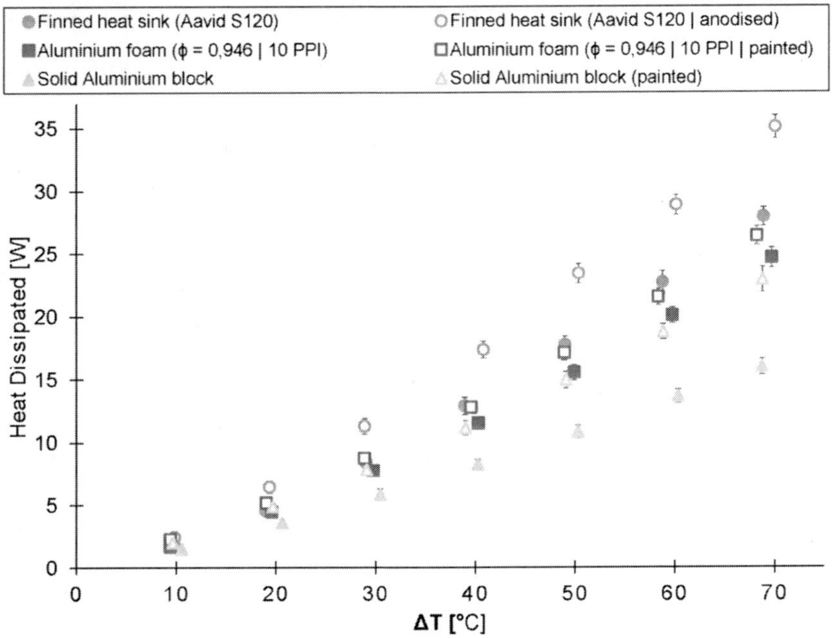

Figure 8. The heat dissipated by the untreated and the painted/anodized heat sink as a function of the temperature difference between the ambient air and the heat sink's base plate (heat sink's height = 22 mm).

If the temperature difference is below 40 ∘ C, no significant difference exists between the heat dissipated by the finned and the aluminum foam heat sink. However, the heat dissipated by the finned heat sink is on average 17% larger for a temperature difference higher than 40 ∘ C. Both the finned and the aluminum foam heat sink clearly outperform the solid aluminum block, except for the lowest temperature difference measured.

The possible performance gain due to the higher surface-to-volume ratio (Table 1) of the aluminum foam heat sink compared to the finned heat sink is not found in the experimental results due to the higher hydraulic resistance of the aluminum foam heat sink.

The difference in performance between the finned and foamed heat sink (reticulated22) is explained by considering the smoke visualizations in Figure 9. The same technique was also used by Taji et al. [31] to visualize the air flow through a heat sink. In Figure 9, smoke generated by burning incense is used to visualize the air flow through the finned and aluminum foam heat sink. The smoke enters both heat sinks at the same location, i.e., at the side near the base plate. However, the exit location differs. In the finned heat sink, the smoke exits at the center of the heat sink. In the case of the aluminum foam heat sink, the smoke exits at the top of the heat sink before it reaches the heat sink's center. After exiting the aluminum foam heat sink, the smoke deflects to the center to rise due to a chimney effect. Due to this deflection, cold air cannot enter the heat sink at the top; thus, no multi-chimney pattern [8] is formed. Hence, only part of the heat exchanging surface area of the aluminum foam adds to the convective heat transfer. Due to the higher hydraulic resistance of metal foam compared to fins, the heat exchanging surface area of foams is used less efficiently than in finned heat sinks. This explains why the metal foam, even though having a larger surface-to-volume ratio (see Table 1), has a smaller thermal performance than the finned heat sink. This shallow penetration by the air in the metal foam was also reported by De Schampheleire et al. [22], who performed numerical simulations. Moreover, a higher temperature difference induces a larger buoyancy force, but the air flow rate through the metal foam will increase less than the one of the finned heat sink, due to its higher hydraulic resistance. This will lower the local convection coefficient.

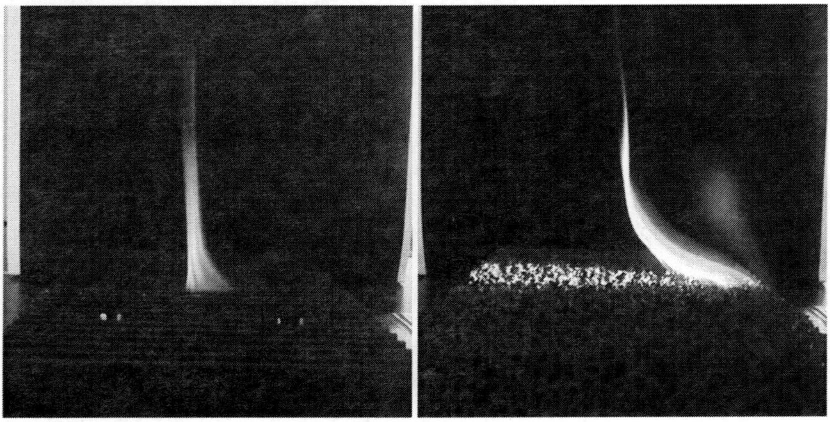

Figure 9. Visualization of the air flow through (a) the finned heat sink (Aavid ® S120) and (b) the in-house-made aluminum foam heat sink (reticulated22) using smoke (ΔT = 70 °C, horizontal orientation).

3.2. Comparison between the Alveotec® and the In-House-Made Aluminum Foam

Figure 10 compares the thermal performance of the in-house-made foam (reticulated40) to that of the Alveotec® foam (spheres40) [25]. The figure shows the heat dissipated by the heat sink as a function of the temperature difference between the heat sink's base plate and the ambient air.

At temperature differences below 50 ∘ C, no significant difference in thermal performance exists between the Alveotec® and the in-house-made foam. Nevertheless, at larger temperature differences, the Alveotec® foam heat sink dissipates up to 12% more heat than the in-house-made aluminum foam heat sink. These results were expected based on the conclusions of previous studies [19,20,21]. These previous studies concluded that the thermal performance of metal foam in buoyancy-induced convection increased with decreasing porosity and PPI-value. A lower PPI-value results in larger pores and, thus, a lower hydraulic resistance, but also a lower surface-to-volume ratio. However, this decrease is partly compensated by the decrease of porosity, which increases the surface-to-volume ratio. Moreover, in buoyancy-induced convection, the hydraulic resistance has to be overcome by a small buoyancy force. Hence, a low hydraulic resistance has probably the most important influence on the thermal performance. Finally, it has to be noted that the in-house-made metal foam was bonded to the base plate using thermal conductive epoxy. This can lead to a higher contact resistance, thereby decreasing the thermal performance [32] according to De Schampheleire et al. [21], the influence of contact resistance is relatively low (<5%) in buoyancy-induced convection.

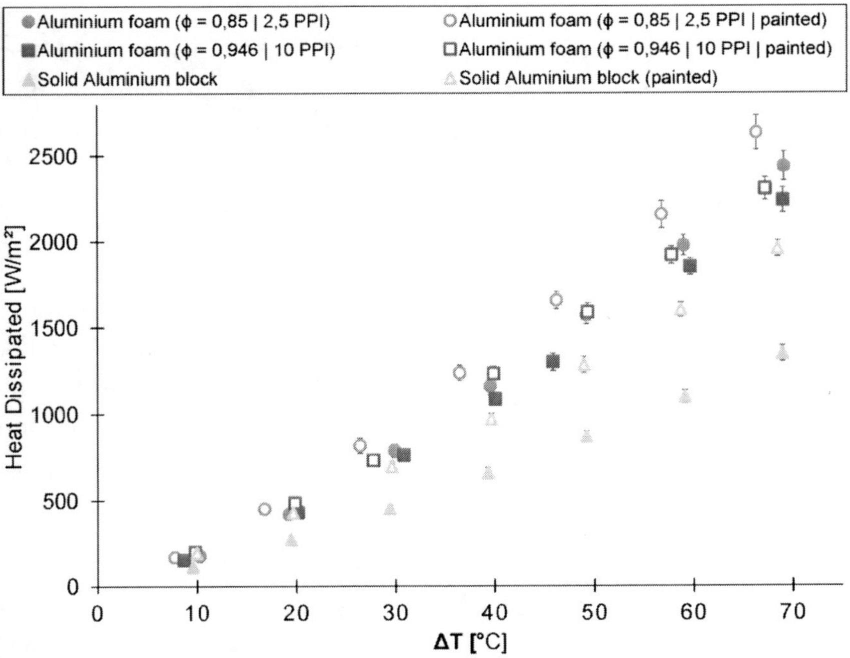

Figure 10. The heat dissipated by the unpainted and painted heat sink as a function of the temperature difference between the ambient air and the heat sink's base plate (heat sink's height = 40 mm).

3.3. Radiative Heat Transfer

To test the influence of thermal radiation on the total heat transfer in buoyancy-induced convection, an untreated and painted/anodized version of each heat sink was tested. Both painting and anodizing increase the emissivity of the heat sink and, consequently, the heat transfer by thermal radiation. The emissivity of the in-house-made aluminum foam was experimentally determined for both the unpainted and painted version using a thermal emissivity measuring instrument (INGLAS® TIR 100-2, Friedrichshafen, Germany). The emissivity of the unpainted and painted in-house-made aluminum foam is, respectively, 0.540 ± 0.005 and 0.894 ± 0.005. The emissivity of the other heat sinks was not tested, but will be of the same order, except for the emissivity of the untreated finned heat sink and the untreated solid aluminum block. The emissivity of those will be much lower than the one of the in-house-made foam, because their surfaces are very smooth. This emissivity lies between 0.05 and 0.25 [33].

Figure 8 compares the thermal performance of the untreated finned heat sink, the 22 mm-high in-house-made aluminum foam heat sink (reticulated22) and the 22 mm-high solid aluminum block to that of their painted/anodized counterparts. By painting or anodizing, the thermal performance improved for each of the tested heat sinks. The aluminum foam heat sink, the finned heat sink and the aluminum block, respectively, dissipated on average 17%, 31% and 44% more heat after they were painted or anodized.

A comparison between the 40 mm-high in-house-made aluminum foam heat sink (reticulated40), the Alveotec® foam heat sink (spheres40) and the 40mm-high solid aluminum block and their painted counterparts is shown in Figure 10. This figure also indicates that the thermal performance of the tested heat sinks in buoyancy-induced convection increases by painting them. The painted version of the 40 mm-high aluminum foam heat sink, the Alveotec® foam heat sink and the 40 mm-high solid aluminum block, respectively, dissipate on average 11%, 21% and 51% more heat than their unpainted versions.

The relative contribution of the thermal radiation decreases with increasing temperature difference, as shown in both Figure 8 and Figure 10. This trend was also found by Sparrow and Vemuri [34] for pin-fin arrays.

Further, the relative contribution of thermal radiation to the total heat transfer decreases with increasing height of the in-house-manufactured aluminum foam heat sink. In contrast; the relative contribution of thermal radiation to the total heat transfer increases with increasing height for the solid aluminum block. According to De Schampheleire et al. [22], the highest aluminum foam has the largest influence by radiation. This conclusion is the opposite of the one found here. The major difference between the experiments of De Schampheleire et al. and the ones of this study are the dimensions of the base plate of the heat sink. The base plate in the study of De Schampheleire et al. [22] is 254 mm by 25.4 mm, whereas in this study, it is 150 mm by 101 mm. The different ratio between the surface area with a large view factor towards the environment and the total surface area could cause the difference in conclusions.

The improvement by painting/anodizing of the finned heat sink is larger compared to the one of the 22 mm-high in-house-manufactured aluminum foam heat sink (Figure 8). To explain this difference in improvement, the Stefan–Boltzmann law is used, which is defined as:

$$\dot{Q}_{\mathrm{rad}} = A\,F\,\epsilon\,\sigma_b\,\left(T_{hs}^4 - T_{amb}^4\right) \qquad 3)$$

Equation (3) calculates the radiative heat transfer towards the environment (Q_{rad}) given the total outer surface area of the subject (A), the view factor (F), the emissivity (ϵ) and the Stefan–Boltzmann constant (σ_b).

The larger improvement by the finned heat sink can partly be explained by its larger increase in emissivity by painting. The increase in emissivity for the finned heat sink is almost twice as large as the one of the aluminum foam. According to Equation (3), the increase of thermal radiation should be higher for the finned heat sink.

Moreover, in the case of the aluminum foam heat sink, the absolute increase in dissipated heat by painting flattens with increasing temperature difference. This is not expected according to the Stefan–Boltzmann law (Equation (3)). This flattening is caused by the lower thermal conductivity of the metal foam itself. The average object's wall temperature, which is an important variable in Equation (3), strongly depends on the thermal conductivity of the heat sink itself. The total heat dissipated by the heat sink increases with increasing temperature difference. This larger heat flux through the heat sink combined with a low thermal conductivity results in a large temperature gradient over the heat sink. This temperature gradient will lower the object's wall temperature and, therefore, the radiative heat transfer. Because of the fourth power in Equation (3), the effect of the wall temperature is more pronounced than for convective heat transfer, reducing the relative contribution of radiative heat transfer to the total heat transfer in buoyancy-induced convection.

3.4. Orientation

Figure 11 represents the thermal performance of a flat plate (bare base plate), a finned heat sink (Aavid® S120) and the in-house-made aluminum foam heat sink, all having the same outer dimensions and all being painted/anodized, in two different orientations: horizontal and vertical. In the vertical orientation, the four-inch side is aligned with gravity.

The aluminum foam heat sink dissipates almost twice as much heat compared to the flat plate at a temperature difference of 70 $\circ$ C. The heat dissipated by the finned heat sink is about three times more than by the flat plate at a temperature difference of 70 ◦ C. For both the flat plate and the finned heat sink, no significant difference in thermal performance can be found between the horizontal and vertical orientation. Nonetheless, the thermal performance of the in-house-made aluminum foam heat sink in the horizontal orientation is up to 18% higher than in the vertical orientation for a temperature difference larger than 30 ◦ C.

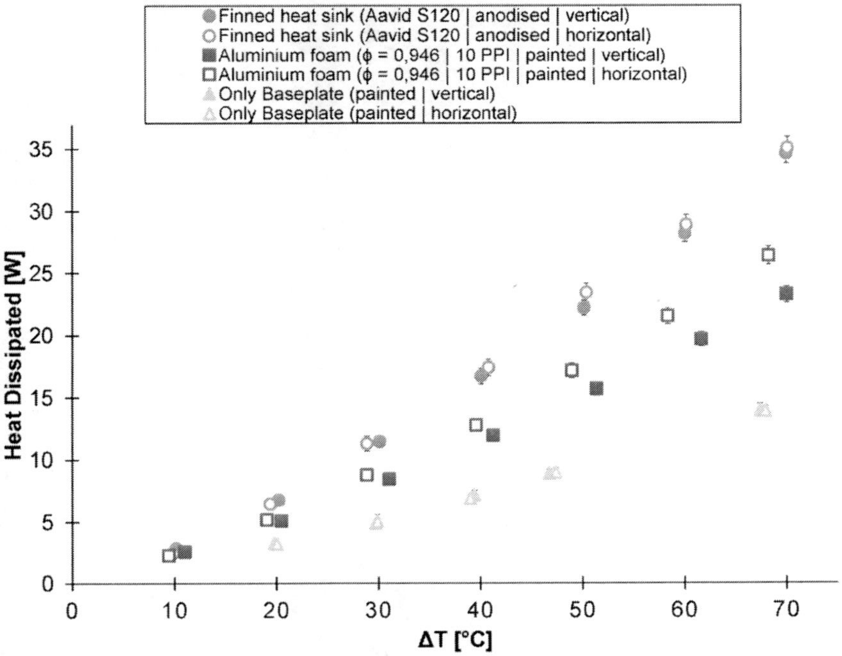

Figure 11. The heat dissipated by the heat sink at different orientations as a function of the temperature difference between the ambient air and the heat sink's base plate (heat sink's height = 22 mm; the 4″-side is parallel with the gravity in the vertical orientation).

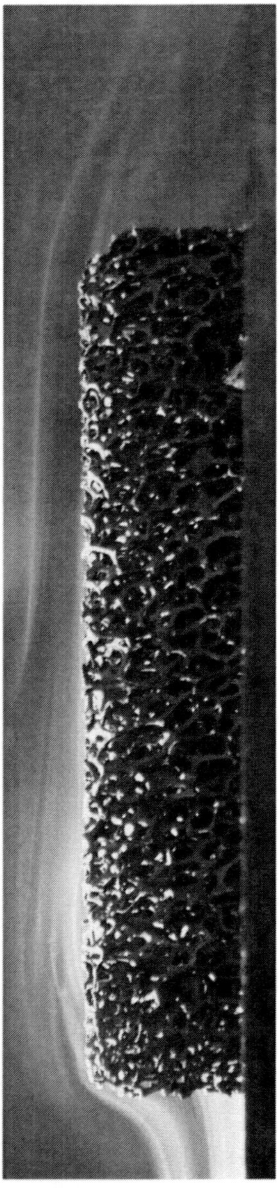

Figure 12. Visualization of the air flow through the in-house-made aluminum foam heat sink (reticulated22) using smoke ($\Delta T = 70$ °C, vertical orientation).

The results obtained in this study are different from those found in the literature [19,20]. Both Qu et al. [20] and Bhattacharya et al. [19] did not find any significant difference between the horizontal and vertical orientation. When this study is compared to the ones of Qu et al. [20] and Bhattacharya et al. [19], it is noted that the temperature test range, foam porosity and pore size are very similar. Yet, the foam used by Qu et al. [20] is made of copper instead of aluminum. Further, the experimental setups and the dimensions of the metal foam heat sinks are different for all three studies. Assuming that the experiments found in the literature are conducted correctly, the only difference that can explain the difference in results is the dimensions of the metal foam heat sink. In contrast to the heat sinks tested in this study, which have a rectangular base plate, the heat sinks tested by Qu et al. [20] have a square base plate of 100 mm by 100 mm. Furthermore, the height of the heat sinks tested by Qu et al. [20] is 55 mm, which is more than double the heat sink's height in this study. No dimensions are specified in the paper of Bhattacharya et al. [19].

Figure 9 and Figure 12 shows that in both horizontal and vertical orientations, the metal foam is penetrated very shallowly by the air flow. This means that in the horizontal orientation, only the regions near the four sides of the foam have a large contribution to the convective heat transfer. Hence, having a rectangular base plate is probably beneficial due to the larger perimeter of the base plate for a given foam volume.

4. CONCLUSIONS

In this study, the thermal performance of seven different aluminum heat sinks in buoyancy-induced convection was tested. All heat sinks had the same base plate dimensions (6″× 4″). During the experiments, the temperature difference between the ambient air and the heat sink's base plate was varied from 10 ∘ C to 70 ∘ C in steps of 10 ∘ C. At each temperature difference, the power dissipated by the heat sink was measured. Based on these experiments, the following conclusions can be made:

- At temperature differences higher than 30 ∘ C, significant differences in thermal performance exist between the finned and metal foam heat sinks at the tested conditions. The finned heat sink dissipates on average 17% more heat than the 22 mm-high in-house-manufactured aluminum foam heat sink. The Alveotec® foam heat sink dissipates 12% more heat than the 40 mm-high in-house-made foam heat sink.
- By increasing the emissivity of a heat sink, the power dissipated by a heat sink in buoyancy-induced convection can be significantly increased by up to 50%. The relative attribution of radiative heat transfer to the total heat transfer decreases for all heat sinks with an increasing temperature difference.
- For the tested conditions, metal foam heat sinks dissipate up to 18% more heat in the horizontal orientation.

ACKNOWLEDGMENTS

Research funded by a Ph.D. grant (No. 141733) of the Agency for Innovation by Science and Technology (IWT).

AUTHOR CONTRIBUTIONS

Marijn Billiet, Sven De Schampheleire and Henk Huisseune conceived and designed the experiments. Marijn Billiet performed the experiments and analyzed the data. Marijn Billiet, Sven De

Schampheleire and Henk Huisseune wrote the paper. Michel De Paepe supervised the study.

REFERENCES

1. Burris, M. Component Failures: How They Fail and How to Identify Failed Components. Available online: http://components.about.com/od/Components/a/Component-Failures.htm (accessed on 30 July 2015).
2. U.S. Department of Energy. Building Technologies Program: Solid-State Lighting Technology Fact Sheet. Available online: http://www.webcitation.org/6aKbWerxo (accessed on 30 July 2015).
3. Meyaard, D.S.; Shan, Q.; Cho, J.; Fred Schubert, E.; Han, S.H.; Kim, M.H.; Sone, C.; Jae Oh, S.; Kyu Kim, J. Temperature dependent efficiency droop in GaInN light-emitting diodes with different current densities. Appl. Phys. Lett. 2012, 100.
4. CREE®; CREE Inc., Durham, NC, USA. Thermal Management of Cree® XLamp® LEDs. Available online: http://www.webcitation.org/6aKchRqUw (accessed on 30 July 2015).
5. Kolar, J.; Drofenik, U.; Biela, J.; Heldwein, M.; Ertl, H.; Friedli, T.; Round, S. PWM Converter Power Density Barriers. In Proceedings of the Power Conversion Conference (PCC), Nagoya, Japan, 2–5 April 2007; pp. 9–29.
6. Pop, E. Energy dissipation and transport in nanoscale devices. Nano Res. 2010, 3, 147–169.
7. Calamas, D.; Baker, J. Behavior of Thermally Radiating Tree-like Fins. Heat Transfer 2013, 135.

8. Dogan, M.; Sivrioglu, M.; Yılmaz, O. Numerical analysis of natural convection and radiation heat transfer from various shaped thin fin-arrays placed on a horizontal plate-a conjugate analysis. Energy Convers. Manage. 2014, 77, 78–88.

9. Bornoff, R.; Parry, J. An additive design heat sink geometry topology identification and optimisation algorithm. In Proceedings of the 31st Thermal Measurement, Modeling Management Symposium (SEMI-THERM), San Jose, CA, USA, 15–19 March 2015; pp. 303–308.

10. De Jaeger, P.; T'Joen, C.; Huisseune, H.; Ameel, B.; de Paepe, M. An experimentally validated and parameterized periodic unit-cell reconstruction of open-cell foams. J. Appl. Phys. 2011, 109.

11. Ashby, F. Metal Foams: A Design Guide; Chemical Petrochemical & Process; Butterworth-Heinemann: Oxford, UK, 2000.

12. Banhart, J. Manufacture, characterisation and application of cellular metals and metal foams. Prog. Mater Sci. 2001, 46, 559–632. [Google Scholar] [CrossRef]

13. Dukhan, N. Metal Foams: Fundamentals and Applications; Destech Publications: Lancaster, PA, USA, 2013.

14. Volz, R. Reticulated Polyurethane Foams and Process for Their Production. U.S. Patent 3,171,820, 2 March 1965.

15. Jinnapat, A.; Kennedy, A. The Manufacture and Characterisation of Aluminum Foams Made by Investment Casting Using Dissolvable Spherical Sodium Chloride Bead Preforms. Metals 2011, 1.

16. Zhao, C. Review on thermal transport in high porosity cellular metal foams with open cells. Int. J. Heat Mass Transfer 2012, 55, 3618–3632.

17. Mancin, S.; Zilio, C.; Diani, A.; Rossetto, L. Air forced convection through metal foams: Experimental results and modeling. Int. J. Heat Mass Transfer 2013, 62, 112–123.

18. De Schampheleire, S.; de Jaeger, P.; Huisseune, H.; Ameel, B.; T'Joen, C.; de Kerpel, K.; de Paepe, M. Thermal hydraulic performance of 10 PPI aluminum foam as alternative for louvered fins in an HVAC heat exchanger. Appl. Therm. Eng. 2013, 51, 371–382.

19. Bhattacharya, A.; Mahajan, R.L. Metal Foam and Finned Metal Foam Heat Sinks for Electronics Cooling in Buoyancy-Induced Convection. J. Electron. Packag. 2005, 128, 259–266.

20. Qu, Z.; Wang, T.; Tao, W.; Lu, T. Experimental study of air natural convection on metallic foam-sintered plate. Int. J. Heat Fluid Flow 2012, 38, 126–132.

21. De Schampheleire, S.; de Jaeger, P.; Reynders, R.; de Kerpel, K.; Ameel, B.; T'Joen, C.; Huisseune, H.; Lecompte, S.; de Paepe, M. Experimental study of buoyancy-driven flow in open-cell aluminum foam heat sinks. Appl. Therm. Eng. 2013, 59, 30–40.

22. De Schampheleire, S.; de Kerpel, K.; de Jaeger, P.; Huisseune, H.; Ameel, B.; de Paepe, M. Buoyancy driven convection in open-cell metal foam using the volume averaging theory. Appl. Therm. Eng. 2015, 79, 225–233.

23. Sikka, K.; Torrance, K.; Scholler, C.; Salanova, P. Heat sinks with fluted and wavy plate fins in natural and low-velocity forced convection. IEEE Trans. Compon. Packag. Technol. 2002, 25, 283–292.

24. Khor, Y.K.; Hung, Y.M.; Lim, B.K. On the role of radiation view factor in thermal performance of straight-fin heat sinks. Int. Commun. Heat Mass Transfer 2010, 37, 1087–1095.

25. Alvéotec; Venissieux, France. Innovation Fonderie. Available online: http://www.alveotec.fr/en/our-news/unique-aluminum-foam-from-goodfellow_41.html (accessed on 24 March 2015).

26. Brun, E. De L'Imagerie 3D des Structures a L'Etude des Mecanismes de Transport en Milieux Cellulaires; Institut Universitaire des Systèmes Thermiques Industriels, Aix-Marseille Université: Marseille, France, 2009.

27. Aavid Thermalloy (Laconia, NH, USA). Available online: http://www.aavid.com/ (accessed on 24 March 2015).

28. Taylor, J. An Introduction to Error Analysis: The Study of Uncertainties in Physical Measurements; University Science Books: Dulles, VI, USA, 1997.

29. Churchill, S.W.; Chu, H.H. Correlating equations for laminar and turbulent free convection from a vertical plate. Int. J. Heat Mass Transfer 1975, 18, 1323–1329.

30. Nellis, G.; Klein, S. Heat Transfer; Cambridge University Press: Cambridge, UK, 2012.

31. Taji, S.; Parishwad, G.; Sane, N. Experimental investigation of heat transfer and flow pattern from heated horizontal rectangular fin array under natural convection. Heat Mass Transfer 2014, 50, 1005–1015.

32. De Jaeger, P.; T'Joen, C.; Huisseune, H.; Ameel, B.; de Schampheleire, S.; de Paepe, M. Assessing the influence of four bonding methods on the thermal contact resistance of open-cell aluminum foam. Int. J. Heat Mass Transfer 2012, 55, 6200–6210.

33. Fluke®; Everett, WA, USA. Emissivity values of common materials. Available online: http://www.webcitation.org/6aKc191IN (accessed on 30 July 2015).

34. Sparrow, E.; Vemuri, S. Orientation effects on natural convection/radiation heat transfer from pin-fin arrays. Int. J. Heat Mass Transfer 1986, 29, 359–368.

6

CHAPTER

COMBINED EFFECTS OF PIPE DIAMETER, REYNOLDS NUMBER AND WALL HEAT FLUX AND ON FLOW, HEAT TRANSFER AND SECOND-LAW CHARACTERISTICS OF LAMINAR-TRANSITIONAL MICRO-PIPE FLOWS

A. Alper Ozalp

Department of Mechanical Engineering, University of Uludag, 16059 Gorukle, Bursa, Turkey

ABSTRACT

Fluid flow, heat transfer and entropy generation characteristics of micro-pipes are investigated computationally by considering the simultaneous effects of pipe diameter, wall heat flux and Reynolds number in detail. Variable fluid property continuity, Navier-Stokes and energy equations are numerically handled for wide ranges of pipe diameter (d = 0.50–1.00 mm), wall heat flux (q '' = 1000–2000 W/m2) and Reynolds number (Re = 1 - 2000), where the relative roughness is kept constant at ε/d = 0.001 in the complete set of the scenarios considered. Computations indicated slight shifts in velocity profiles from the laminar character at Re = 500 with the corresponding shape factor (H) and intermittency values (γ) of H = 3.293→3.275 and γ = 0.041→0.051 (d = 1.00→0.50 mm). Moreover, the onset of transition was determined to move down to Retra = 1,656, 1,607, 1,491, 1,341 and 1,272 at d = 1.00, 0.90, 0.75, 0.60 and 0.50 mm, respectively. The impacts of pipe diameter on friction mechanism and heat transfer rates are evaluated to become more significant at high Reynolds numbers, resulting in the rise of energy loss data at the identical conditions as well. In cases with low pipe diameter and high Reynolds number, wall heat flux is determined to promote the magnitude of local thermal entropy generation rates. Local Bejan numbers are inspected to rise with wall heat flux at high Reynolds numbers, indicating that the elevating role of wall heat flux on local thermal entropy generation is dominant to the suppressing function of Reynolds number on local thermal entropy generation. Cross-sectional total entropy generation is computed to be most influenced by pipe diameter at high wall heat flux and low Reynolds numbers.

Keywords: micro-pipe; friction coefficient; heat transfer; entropy generation

1. INTRODUCTION

With the progress of scientific investigations on energy management by developing new power supervision strategies, aiming to increase efficiency values and to suppress power loss amounts, the focus on this research branch is mainly directed towards the design and integration of micro-systems in actual industrial processes. The simultaneous advance of scientific efforts and the related manufacturing technologies recently made it possible for the micro-flow applications to take part in various engineering fields such as heat exchangers [1], fuel cells [2], pumps [3] and even in gas turbines [4]. Showing parallelism, both research outputs and industrial feedback signified the importance of surface roughness in micro-pipe flows by associating the roughness based viscous effects on the development of velocity and temperature profiles, flow-transition activity, entropy generation and power loss rates. Thus, due to these scientific fundamentals and industrial necessities, a research study, focusing on the roles of surface roughness in micro-pipe flows, must be structured on the three main issues: (i) frictional activity and flow characteristics, (ii) heat transfer mechanism and (iii) second-law analysis.

The roles of surface roughness on the friction characteristics, flow behavior and transitional activity in micro-pipe flows have been in the agenda of several researchers and there exist several published experimental and numerical reports. The role of frictional activity on the early transition from laminar to turbulent flow and on the grow of surface friction with surface roughness was reported by Guo and Li [5]. Celata et al. [6] determined, for the micro-pipe diameter range of $d < 100$ μm with the non-dimensional roughness of $\varepsilon^* < 0.01$, that friction factor tends to deviate from the Poiseuille law when $Re > 1300$, which can be attributed to the acceleration associated with compressibility effect. As energy conversion of near-wall microfluidic transport was numerically investigated for slip-flow conditions, including different channel aspect ratios, pressure coefficients and slip flow conditions, by Ogedengbe et al. [7], Engin et al. [8] studied wall roughness effects in microtube flows and evaluated considerable shifts from the conventional laminar flow theory. Laminar and transitional flow experiments in dimpled tubes were performed by Vicente et al. [9], who reported a relatively low transition Reynolds number of 1,400 where the roughness induced friction factors were 10% higher than those of the smooth tubes. In a similar work, Kandlikar et al. [10] studied surface roughness effects on pressure drop in circular small hydraulic diameter tubes, where the transition onset appeared at Reynolds number values much below 2,300. Wen et al. [11] experimentally investigated the augmentation of pressure drop characteristics due to different strip-type inserts in small tubes. As Koo and Kleinstreuer [12] numerically and experimentally studied the significance of viscous dissipation on the friction factor, the roles of surface roughness on viscous dissipation, increased friction factor values and the corresponding earlier transitional activity was expressed by Celata et al. [13,14]. Morini [15] handled the concept of viscous dissipation for various duct cross-section types, and pointed out the minimum Reynolds number for which viscous dissipation effects can not be neglected. A comprehensive review, on the interaction of surface roughness with frictional activity and momentum rates in microchannels, was carried out by Obot [16]. He reported that a wide collection of available literature identified the definition of microchannels with a hydraulic diameter of dh ≤ 1,000 μm (1 mm), moreover, the limiting Reynolds number for the onset of transition in smooth microchannels was above 1,000.

The role of surface roughness in thermal systems was also taken into consideration in recent studies; some dealt with the roughness definition and others focused on the heat transfer mechanism. Wang et al. [17] investigated the influence of two-dimensional roughness on laminar flow in microchannels between two parallel plates by applying the regular perturbation method. Sheikh et al. [18] developed a model to eliminate the discrepancy in the fouling measurements by characterizing fouling as a correlated random process. In a complementary work, the role of fouling on entropy generation, as compared to that for clean surface tubes, was studied by Sahin et al. [19]. Cao et al. [20] employed non-equilibrium molecular dynamics simulation to investigate the effect of the surface roughness on slip flow of gaseous argon. More comprehensive research comprised both flow and heat transfer characteristics driven by roughness, such as Wu and Cheng [6], who reported on the Nusselt number augmentations with surface roughness, Kandlikar et al. [10] findings on heat transfer manipulation due to roughness, Wen et al. [11] work on the effects of the imposed wall heat flux, mass flux and strip inserts on heat transfer rates in small diameter tubes, Koo and Kleinstreuer's [12] evaluations on the temperature field with viscous dissipation, a review by Obot [16] indicating the variation of laminar Nusselt numbers with the square root of the Reynolds number and finally Ozalp's [21,22] numerical compressible flow investigations for aerospace propulsion applications.

Since the essential concern of thermo-fluid based research studies is to generate a complete overview on the schematic efficiency features, second-law analysis recently became prominent due to its potential to describe the overall system performance. Kotas et al. [23] indicated that for calculating loss of exergy, or process irreversibility, an exergy balance or the Gouy-Stodola theorem can be used. Zimparov [24] reviewed passive heat transfer augmentation techniques for single-phase flows. The effects of the spanwise fin tube spacing on the second-law performance of wavy plate fin-and-tube heat exchangers were investigated by Lin and Lee [25]. Ratts and Raut [26] obtained optimal Reynolds numbers for single-phase, convective, fully developed internal laminar and turbulent flows with uniform heat flux, by employing the entropy generation minimization method. Entropy generation for a fully developed laminar viscous flow in a duct subjected to constant wall temperature was analytically studied by Sahin [27]. His computations not only show that temperature dependency of viscosity becomes essentially important in accurately determining the entropy generation, but also indicate that for low heat transfer conditions the entropy generation due to viscous friction becomes dominant. Rakshit and Balaji [28] reported the results of a numerical investigation of conjugate convection from a finned channel, where the vertical rectangular fins were mounted outside the horizontal channel. The effects of streamwise variation of fluid temperature and rib height to diameter ratio on the entropy production of a tubular heat exchanger with enhanced heat transfer surfaces were investigated by Zimparov [29]. In singly connected microchannels with finite temperature differences and fluid friction, Richardson et al. [30] seeked the optimum laminar flow regime based on second-law analysis. Ko [31] computationally carried out the thermal design of plate heat exchanger double-sine ducts, from the point of entropy generation and exergy utilization. Laminar forced convection and entropy generation in a helical coil with constant wall heat flux was also numerically investigated by Ko [32]. He suggested that the optimal Reynolds number is to be chosen according to the flow operating condition, for least thermal irreversibility and best exergy utilization

The past research categorized above clearly indicates the wide scientific prospects and the industrial importance and actuality of the micro-pipe flows with surface roughness. As the available literature designates, real-time operational and structural parameter ranges in micro-pipe flows are broad, which as a consequence significantly promotes the possible process and output scenarios due to the individual and combined parameter influences. Recently, Ozalp [33,34,35] computationally investigated the momentum transfer, heat transfer and second-law characteristics of air flow in a circular micro-pipe with a diameter of d = 1 mm. The analyses were carried out in the Reynolds number and non-dimensional surface roughness ranges of Re = 1–2,000, ε^* = 0.001–0.05. The author continued his research by expanding the micro-pipe diameter and non-dimensional surface roughness limits to d = 0.10–1.00 mm and ε^* = 0.001–0.1 for water flow; this content is recently supported by the Uludag University Research Fund. In the present article, the author reports the initial finding set of the ongoing project, comprising the first and second-law characteristics of roughness induced laminar-transitional micro-pipe flow, for the diameter range of d = 0.50–1.00 mm with the fixed non-dimensional surface roughness condition of ε* = 0.001. Radial velocity and temperature profiles, boundary layer parameters, friction coefficients, Nusselt numbers, frictional energy loss and temperature rise values are discussed to identify the flow and heat transfer characteristics. Additionally, second-law records are interpreted through radial profiles of thermal and frictional entropy values and cross-sectional total entropy generation rates and local and cross-sectional average Bejan numbers.

2. THEORETICAL BACKGROUND

2.1. Micro-Pipe and Roughness

The diameter and length of the micro-pipe (Figure 1a) analyzed in the present paper, are given as d and L, respectively. The present roughness model is based on the triangular structure of Cao et al. [20] (Figure 1b), where the roughness amplitude is characterized by ε and period by ω. In all computations, the roughness periodicity parameter (ω' = ω/ε) is kept fixed to ω' = 2.31, which corresponds to the equilateral triangle structure (Cao et al. [20]).

Equation 1 numerically characterizes the model of Cao et al. [20] with the implementation of the amplitude and period. The model function (fε(z)) is repeated in the streamwise direction throughout the pipe length, where the Kronecker unit tensor (δi) attains the values of δI = +1 and –1 for 0 ≤ z ≤ 2.312 ε and 2.312 ε ≤ z ≤ 2.31ε respectively.

$$f_\varepsilon(z) = \delta_i \varepsilon \left[1 - \frac{4}{2.31\varepsilon} z \right] \tag{1}$$

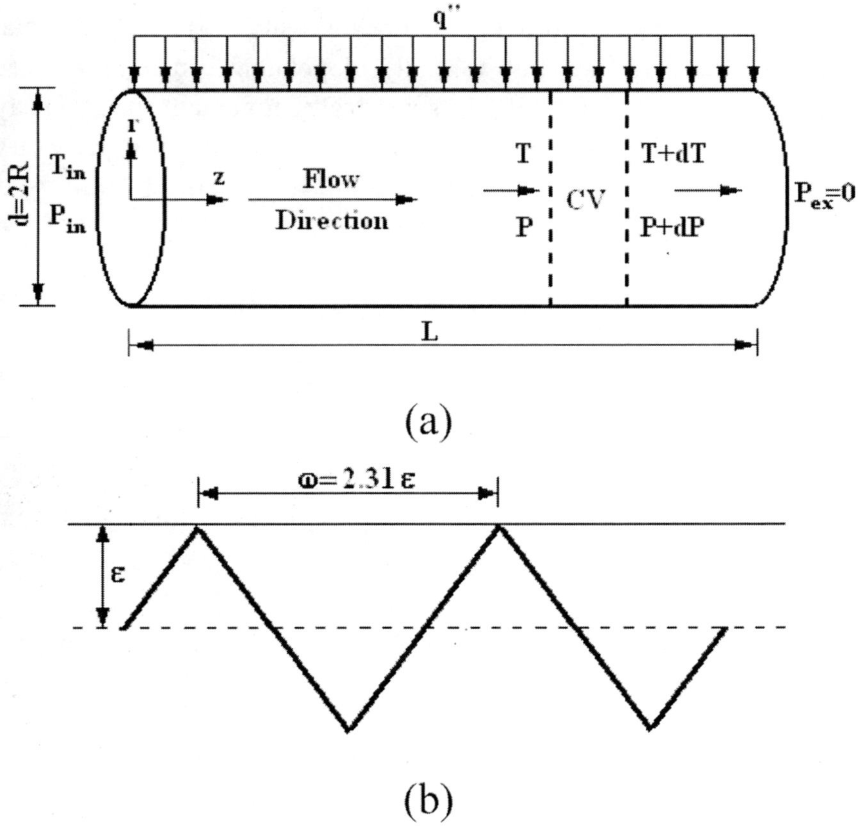

Figure 1. (a) Schematic view of micro-pipe, (b) triangular surface roughness distribution.

2.2. Governing Equations

The density, dynamic and kinematic viscosity of water are defined with ρ, μ and ν (=μ/ρ); the average velocity of the flow at any cross-section of the duct is denoted by Uo. As the surface and mean flow temperatures are denoted by Ts and To, thermal conductivity and convective heat transfer coefficient are characterized by κf and h. Using these definitions, Reynolds number and Nusselt number are given by Equations 2a-b. Water has been selected as the working fluid in the present study, and it is well known that water properties like specific heat (Cp), kinematic viscosity (ν) and thermal conductivity (κf), are substantially dependent on temperature (Incropera and DeWitt [36]). To sensitively implement the property (ξ) variations with temperature into the calculations, the necessary water data of Incropera and De Witt [36] are fitted into 6th order polynomials, which can be presented in closed form by Equation 2c. The uncertainty of the fitted water data is less than 0.03% and the temperature dependency is indicated by the superscript T throughout the formulation.

$$Re = \frac{U_o d}{\nu^T} = \frac{\rho U_o d}{\mu^T} \qquad Nu = \frac{hd}{\kappa_f^T} = \frac{\partial T / \partial r \big|_{r=R} \, d}{T_s - T_o} \qquad \zeta^T = \sum_{j=0}^{6} a_j T^j \tag{2a-c}$$

The problem considered here is steady (∂ ∂t /)=0 , fully developed and the flow direction is coaxial with pipe centerline (U r =U θ =0) , thus the velocity vector simplifies to V⃗ =U z (r)k^ , denoting ∂U z ∂z / =0 and ∂U z ∂θ / =0 . These justifications are common in several recent numerical studies on roughness induced flow and heat transfer investigations, like those of Engin et al. [8], Koo and Kleinstreuer [12] and Cao et al. [20]. The flow boundary conditions are based on the facts that on the pipe wall no-slip condition and constant heat flux exist, and flow and thermal values are maximum at the centerline. As given in Figure 1a, , pressure and temperature values are known at the pipe inlet and the exit pressure is atmospheric. Denoting Uz=Uz(r) and T=T(r,z), the boundary conditions can be summarized as follows:

$$r=R+f_\varepsilon(z) \;\rightarrow\; U_z=0 \quad \& \quad r=0 \;\rightarrow\; \frac{\partial U_z}{\partial r}=0$$

$$r=R+f_\varepsilon(z) \;\rightarrow\; \frac{\partial T}{\partial r} = -\frac{q_r''}{\kappa_f^T} \quad \& \quad r=0 \;\rightarrow\; \frac{\partial T}{\partial r}=0 \tag{3}$$

Since the viscous stress τzz is characterized by the variation of flow velocity in the axial direction (∂U z /∂z) , in an incompressible flow application with constant pipe diameter, it vanishes due to the unvarying local and cross-sectional average velocities in the flow direction. Thus, as the components of the viscous stress tensor (τrz) and heat flux terms (q '' r ,q '' z) can be written as Equations 4a-c, internal and kinetic energy terms are defined as e = CpT and k = U 2 z /2, respectively.

$$\tau_{rz} = \mu^{T} \frac{\partial U_z}{\partial r} \qquad q_r'' = -\kappa_f^{T} \frac{\partial T}{\partial r} \qquad q_z'' = -\kappa_f^{T} \frac{\partial T}{\partial z} \tag{4a-c}$$

Application of wall heat flux (q '') on the lateral surface of the micro-pipe results in the total surface heat transfer of Qs = q '' πdL. With these problem definitions and above implementations for a laminar incompressible flow with variable fluid properties, continuity, momentum and energy equations are given as:

$$\frac{\partial}{\partial z}(U_z) = 0 \tag{5}$$

$$\frac{\partial P}{\partial z} = \frac{1}{r} \frac{\partial}{\partial r}(r\tau_{rz}) \tag{6}$$

$$\rho U_z \frac{\partial}{\partial z}\left(e + \frac{P}{\rho} + k \right) + \frac{1}{r}\frac{\partial}{\partial r}(r q_r'') + \frac{\partial q_z''}{\partial z} = \tau_{rz} \frac{\partial U_z}{\partial r} + U_z \left[\frac{1}{r}\frac{\partial}{\partial r}(r\tau_{rz}) \right] \tag{7}$$

The cross-sectional average fluid velocity and temperature are obtained from

$$U_o = \frac{2\pi \int_{r=0}^{r=R} U_z(r) r dr}{\pi R^2} \qquad T_o = \frac{2\pi \int_{r=0}^{r=R} U_z(r) C_p(r) T(r) r dr}{U_o (C_p)_o \pi R^2} \tag{8a-b}$$

and the shear stress and mass flow rate are defined as

$$\tau = C_f \frac{1}{2} \rho U_o^2 = \mu^{T} \left| \frac{dU_z}{dr} \right|_{r=R} \qquad \dot{m} = \rho U_o A = \rho 2\pi \int_{r=0}^{r=R} U_z(r) r dr \tag{9a-b}$$

Gouy-Stodola theorem [23] considers the lost available work to be directly proportional to entropy generation, which results from the non-equilibrium phenomenon of exchange of energy and momentum within the fluid and at the solid boundaries. Due to the existence of the velocity and temperature gradients in the flow volume, the volumetric entropy generation rate is positive and finite. Computation of the temperature and the velocity fields through Equations 5–7 on the problem domain will produce the input data for Equation 10a. Equation 10a defines the local rate of entropy generation per unit volume (S'''), for a one-dimensional flow and two-dimensional temperature domain for incompressible Newtonian fluid flow in cylindrical coordinates.

$$S''' = \frac{\kappa_f^T}{T^2}\left[\left(\frac{\partial T}{\partial r}\right)^2 + \left(\frac{\partial T}{\partial z}\right)^2\right] + \frac{\mu^T}{T}\left[\left(\frac{\partial U_z}{\partial r}\right)^2\right] \qquad S''' = S'''_{\Delta T} + S'''_{\Delta P} \tag{10a-b}$$

$$S' = 2\pi \int_{r=0}^{r=R} S''' r dr \qquad Be''' = \frac{S'''_{\Delta T}}{S'''} \qquad Be = \frac{2\pi \int_{r=0}^{r=R} Be''' r dr}{\pi R^2} \tag{11a-c}$$

As given in Equation 10b, the entropy generation due to finite temperature differences ($S'''_{\Delta T}$) in axial z and in radial r directions is defined by the first term on the right side of Equation 10a, the second term defines the frictional entropy generation ($S'''_{\Delta P}$). Equation 11a stands for the cross-sectional total entropy generation rate; local and cross-sectional average values of Bejan numbers can be obtained by Equations 11b-c.

2.3. Computational Method

The author employed the present computational method in his former work [22,33,34,35] and detailed the essentials and the sub-steps of the procedure extensively. Thus, here only the fundamental issues of the methodological structure are defined. To simultaneous handle the continuity, momentum and energy mechanisms of the flow, Equations 5–7 are assembled into the three-dimensional

"Transfer Matrix", consisting of the converted explicit forms of the principle equations. The flow domain of Figure 1a is divided into m axial and n radial cells (m × n), where the fineness of the computational grids is examined to ensure that the obtained solutions are independent of the grid employed. After a series of successive runs the optimum axial and radial cell numbers are determined as m = 500→850 and n = 100→225 respectively, for d = 1.00→0.50 mm. Forward difference discretization is applied in the axial and radial directions in the two-dimensional marching procedure. To sensitively compute the velocity and temperature gradients on the pipe walls, the 20% of the radial region, neighboring solid wall, is employed an adaptive meshing with radial-mesh width aspect ratio of 1.1→1.05 (d = 1.00→0.50 mm). The influences of surface roughness and surface heat flux conditions, over the meshing intervals of the flow domain, are coupled by Direct Simulation Monte Carlo (DSMC) method. The author previously applied DMSC to compressible nozzle flow problems [21,22] and to micro-pipe flow scenarios with surface roughness [33,34,35]. The concept of triple transport conservation is as well incorporated into the DSMC, which makes it possible to sensitively evaluate the balance of heat swept from the micro-pipe walls, the energy transferred in the flow direction and also to perform accurate simulation for inlet/exit pressure boundaries. The "Transfer Matrix" scheme and the DSMC algorithm are supported by cell-by-cell transport tracing technique to guarantee mass conservation, boundary pressure matching and thermal equilibrium within the complete mesh. The resulting nonlinear system of equations is solved by using the Newton-Raphson method and in the case that the Reynolds number does not fit the required value, inlet pressure and temperature are both modified by DSMC, to regulate the Reynolds number of the former iteration step.

3. RESULTS AND DISCUSSION

To generate a scientifically detailed and academically enlightening scheme for the combined effects of pipe diameter, Reynolds number and wall heat flux on the flow, heat transfer and second-law characteristics of laminar-transitional micro-pipe flows, computations are carried out in the ranges of d = 0.50–1.00 mm, Re = 1–2,000 and q ″ = 1,000–2,000 W/m2. As the considered micro-pipe diameter range is consistent with the microchannel definition of Obot [16] (d ≤ 1.00 mm), the imposed wall heat flux values are decided in conjunction with the Reynolds number and the accompanying mass flow rate ranges, to bring about applicable and rational heating. The fixed parameters of the analyses are the length of the micro-pipe (L = 0.5 m), inlet temperature (Tin = 278 K) and exit pressure (Pex = 0 Pa) of water and the non-dimensional surface roughness (ε* = 0.001), which is in harmony with those of Engin et al. [8] (ε* ≤ 0.08), Sahin et al. [19] (ε* ≤ 0.25), Ozalp [33,34,35] (ε* = 0.001–0.05) and Vijayalakshmi et al. [37] (ε* = 0.0004–0.0018). Flow and heat transfer characteristics are presented in terms of radial velocity and temperature profiles, boundary layer parameters, friction coefficients, Nusselt numbers, frictional energy loss and temperature rise values. Radial profiles of thermal and frictional entropy values, cross-sectional total entropy generation rates and Bejan numbers are discussed and identified to interpret the second-law records.

3.1. Analyses of Flow and Heat Transfer Characteristics

3.1.1. Radial schemes

Figure 2 displays the velocity profile (VP) formation in the Reynolds number range of Re = 100–2,000, for the pipe diameters d = 1.00, 0.75 and 0.50 mm with the fixed non-dimensional surface roughness of ε* = 0.001. To clearly display the deviations of the VPs from the characteristic styles of laminar and turbulent regimes and also to put forward the roles of d and Re on the transition mechanism, the characteristic laminar profile (Equation 12a) and the modified turbulent logarithm law for roughness (Equation 12b) (White [38]) are also displayed in Figure 2, where U* is the friction velocity (Equation 12c).

$$\frac{U_z(r)}{U_o} = 2\left[1-\left(\frac{r}{R}\right)^2\right] \qquad \frac{U_z(r)}{U^*} = 2.44\ln\left(\frac{R-r}{\varepsilon}\right)+8.5 \qquad U^* = \sqrt{\frac{\tau_w}{\rho}}$$

(12a-c)

To clarify the roles of micro-pipe diameter and Reynolds number on the frictional activity, the classical and normalized friction coefficient values are evaluated by Equations 13a-c [38]. To strengthen the discussions on transition, boundary layer parameters like shape factor (H) and intermittency (γ) [38] are estimated by Equations 14a and b, respectively.

$$C_f = \frac{2\mu^T \left|\left.\frac{dU_z}{dr}\right|_{r=R}\right|}{\rho U_o^2} \qquad (C_f)_{lam} = \frac{16}{Re} \qquad C_f^* = \frac{C_f}{(C_f)_{lam}}$$

(13a-c)

$$H = \frac{\int_{r=0}^{r=R}\left(1-\frac{U_z(r)}{U_c}\right) r\,dr}{\int_{r=0}^{r=R}\frac{U(r)}{U_m}\left(1-\frac{U_z(r)}{U_c}\right) r\,dr} \qquad \gamma = \frac{H_{lam}-H}{H_{lam}-H_{turb}}$$

(14a-b)

As the laminar (Hlam = 3.36) and turbulent (Hturb = 1.70) shape factor values are computed with Equation 14a, by integrating the laminar (Equation 12a) and turbulent (Equation 12b) profiles, the shape factor data of the transitional flows were also calculated with Equation 14a, however with the computationally evaluated corresponding velocity profiles. The computed boundary layer and transition parameters are presented in Table 1.

Table 1. Boundary layer parameters for various Re and d cases (*: Transitional Reynolds numbers).

d = 1.00 mm				
Re	U_c/U_o	C_f	H	γ
100	1.993	0.1609	3.348	0.007
500	1.960	0.0329	3.293	0.041
1000	1.919	0.0170	3.223	0.082
1500	1.879	0.0116	3.159	0.121
*1656**	*1.867*	*0.0106*	*3.133*	*0.137*
2000	1.838	0.0090	3.097	0.158

d = 0.90 mm				
Re	U_c/U_o	C_f	H	γ
100	1.992	0.1609	3.348	0.007
500	1.959	0.0330	3.291	0.042
1000	1.917	0.0170	3.219	0.085
1500	1.875	0.0117	3.153	0.125
*1607**	*1.865*	*0.0110*	*3.136*	*0.135*
2000	1.833	0.0090	3.090	0.163

d = 0.75 mm				
Re	U_c/U_o	C_f	H	γ
100	1.992	0.1610	3.347	0.008
500	1.957	0.0330	3.286	0.044
1000	1.912	0.0171	3.211	0.090
*1491**	*1.868*	*0.0115*	*3.144*	*0.130*
1500	1.867	0.0117	3.141	0.132
2000	1.823	0.0091	3.075	0.172

d = 0.60 mm				
Re	U_c/U_o	C_f	H	γ
100	1.991	0.1610	3.346	0.008
500	1.953	0.0331	3.280	0.048
1000	1.905	0.0171	3.200	0.097
*1341**	*1.871*	*0.0131*	*3.154*	*0.124*
1500	1.857	0.0118	3.125	0.141
2000	1.809	0.0091	3.056	0.183

d = 0.50 mm				
Re	U_c/U_o	C_f	H	γ
100	1.991	0.1611	3.345	0.009
500	1.950	0.0332	3.275	0.051
1000	1.899	0.0172	3.190	0.102
*1272**	*1.870*	*0.0138*	*3.159*	*0.121*
1500	1.848	0.0119	3.112	0.150
2000	1.797	0.0092	3.039	0.193

It can be inspected from Figure 2 that in the all pipe diameter cases investigated, the VPs for the Re = 100 flow fit with the laminar profile, where the shape factor values are evaluated as H ≥ 3.345 (Table 1) being very close to the characteristic laminar value of H = 3.36. In the Re = 500 case, the VPs are still in harmony with the laminar profile; however slight shifts from the laminar character can be detected, such that the center velocity values and the shape factor and intermittency data build up the ranges of Uc/Uo = 1.960→1.950, H = 3.293→3.275 and γ = 0.041→0.051 (d = 1.00→0.50 mm), respectively.

Moreover, in the scenarios with higher Re, recognizable deviations in the VPs are determined. Thus to identify the modifications, the near wall (0.998 ≤ r/R ≤ 1) and pipe-centerline (0 ≤ r/R ≤ 0.4) regions are demonstrated through zoomed plots. Computations put forward that in the systems with higher Re and/or lower d as Cf elevates, the VPs depart significantly from the laminar profile causing the center velocity figures (Uc/Uo) to decrease, where these evaluations can also be visualized from the zoomed plots and Table 1. Since friction coefficient is mainly characterized by the rate of viscous momentum exchange at the solid boundary (C f =μ T |dU z dr / | r=R /1 2 / ρU 2 o) , the encouraged advance of the wall velocity gradient ($|dU_z/dr|_{r=R}$) with higher Re and/or lower d can be identified as the primary flow catalyst motivating the grow of Cf (Table 1). The near wall ($0.998 \leq r/R \leq 1$) zoomed plots of Figure 2 clarify that for flows with lower micro-pipe diameter the impact of Reynolds number on VPs is more noticeable. This evaluation is significant to interpret the combined effects of Re and d on the momentum transfer and transition mechanism in micro-pipe flows. In the light of several numerical and experimental research studies that examine the roles of micro-pipe diameter and surface roughness on the transition mechanism in internal flow problems, many of the involved scientists [5,9,10,16,39] accept the 10% rise of Cf ($Cf^* = 1.1$) above the laminar value (Equation 13b) as an indicator for the transitional onset. Analyzing the present friction coefficient findings with this method sensitively mark the transitional Reynolds numbers for the micro-pipe diameters of d = 1.00, 0.90, 0.75, 0.60 and 0.50 mm as Re_{tra} = 1,656, 1,607, 1,491, 1,341 and 1,272 respectively. These numbers point out that the structural inducement (d = 1.00→0.50 mm) of the micro flow character with the encouraged transition behavior at relatively lower Re. The open literature proposes a number of transitional micro-pipe studies. Although the common point is the encouraged transitional activity at lower micro-pipe diameters, the findings on transitional Reynolds numbers do mostly not closely resemble but interpret a general attitude. Such as: Guo and Li [5] Re_{tra} = 1800 for d = 0.13–0.18 mm, Celata et al. [13] Re_{tra} = 100 for d = 0.13 mm, Ozalp [35] Re_{tra} = 1650→450 (ε^* = 0.001→0.05) for d = 1.00 mm, Wu and Little [39] Re_{tra} = 400–900 for d=0.05–0.08 mm, Yu et al. [40] Re_{tra} = 1700 and Choi et al. [41] Re_{tra} = 500 for d = 0.01 mm. The basis of this divergence may be due to the undefined/unavailable surface roughness information in many of the works. However, the present findings on transition onset still not only conform with the associated reports but also enriches the related research frame with scientific outputs on well defined scenarios. It can additionally be extracted from the present Re_{tra} that, as the gap of $(Re_{tra})_{d=1.00}$

mm - (Retra)d=0.90 mm is 49, the corresponding one for (Retra)d=0.60 mm - (Retra)d=0.50 mm becomes 69. These values clearly pronounce that, the lower the micro-pipe diameter is reduced the more the sensed influence on Retra is promoted. The tabulated outputs in Table 1 further denote that the transitional shape factor and intermittency are located in the ranges of H = 3.133→3.159 and γ = 0.137→0.121 for d = 1.00→0.50 mm. These transition indicators expose the rise of shape factor values and drop of intermittency data in micro-pipes with lower diameters. Numerical works of Ozalp [34,35] for air flow in a micro-pipe with d = 1.00 mm and ε^* = 0.001 concluded with the transitional shape factor and intermittency outputs of H = 3.141 and γ = 0.132, respectively. Moreover, the zoomed plots ($0 \leq r/R \leq 0.4$) of Figure 2 confirm for the velocity profiles, matching the Retra range of Table 1, that the center velocity values (Uc) generate the scene of $1.85 \leq Uc/Uo \leq 1.90$, which can also be considered as a trustable indicator for transitional activity.

The combined roles of Reynolds number, surface heat flux and micro-pipe diameter on the temperature profile (TP) development are shown in Figure 3. TPs are presented in non-dimensional form with respect to the centerline value (Tc). To identify the considerable influence of viscous dissipation based energy loss $\left(\Psi_{loss} = 2\pi \int_0^R U_z \frac{1}{r}\frac{\partial}{\partial r}(r\tau_{rz})rdrL\right)$ on the TP structure in micro-pipe flows TPs are also compared with the laminar Constant Heat Flux (CHF) formula of Equation 15 [36]. Moreover, the findings on the thermal behavior of micro-pipe flows are supported with the tabulated records of Ψloss and flow temperature rise (ΔT) in Table 2.

$$T(r) = T_s - \frac{2U_o\kappa_f^T R^2}{\rho C_p^T}\left(\frac{dT_o}{dz}\right)\left[\frac{3}{16} + \frac{1}{16}\left(\frac{r}{R}\right)^4 - \frac{1}{4}\left(\frac{r}{R}\right)^2\right] \qquad (15)$$

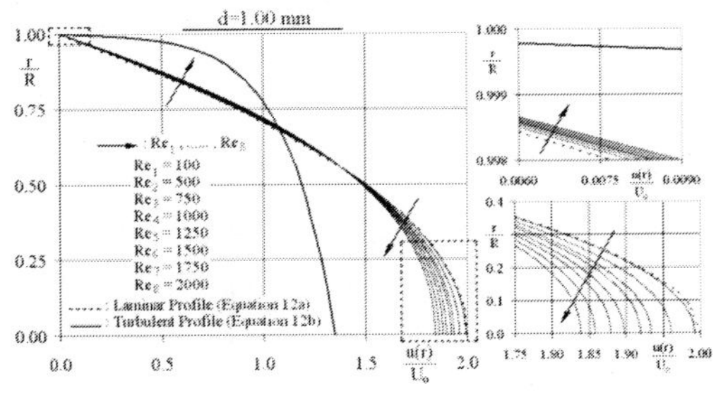

(a)

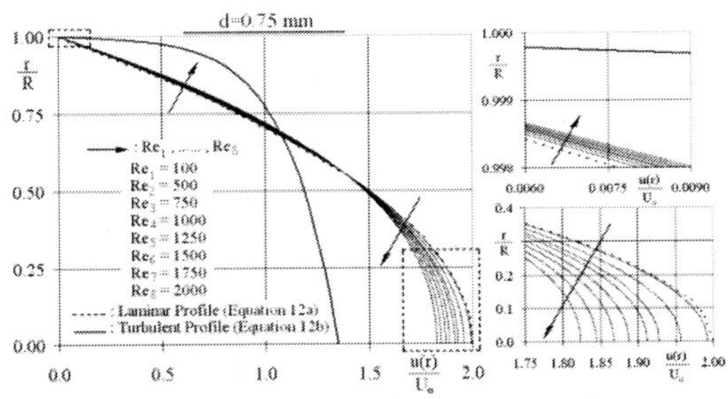

(b)

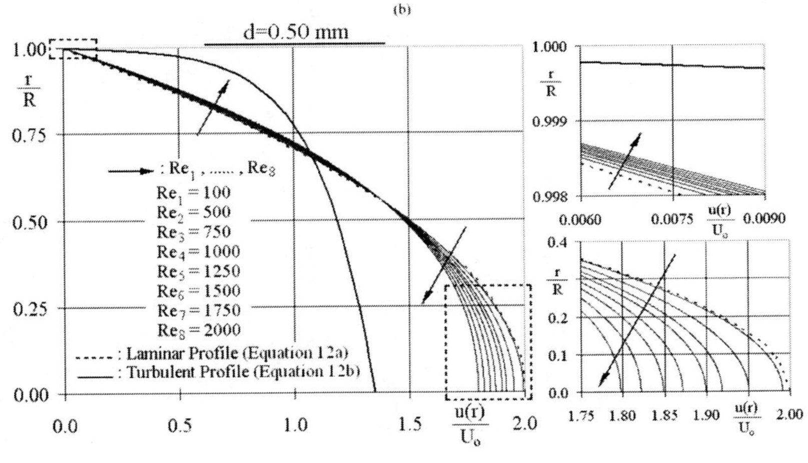

(c)

Figure 2. Variations of radial velocity profiles with Re for the micro-pipe diameters of (a) d = 1.00 mm, (b) d = 0.75 mm and (c) d = 0.50 mm.

Table 2. ΔT and Ψloss values for various Re, d and q '' cases.

d (mm)	q'' (W/m^2)	Q_s (W)	Re = 100		Re = 500		Re = 1000		Re = 1500		Re = 2000	
			ΔT (K)	Ψ_{loss} (W)	ΔT (K)	Ψ_{loss} (W)	ΔT (K)	Ψ_{loss} (W)	ΔT (K)	Ψ_{loss} (W)	ΔT (K)	Ψ_{loss} (W)
1.00	1000	1.571	4.690	1.333E-04	0.936	3.336E-03	0.465	1.337E-02	0.307	3.017E-02	0.227	5.386E-02
	1500	2.356	7.036	1.292E-04	1.405	3.312E-03	0.700	1.332E-02	0.463	3.010E-02	0.344	5.377E-02
	2000	3.142	9.381	1.257E-04	1.874	3.289E-03	0.934	1.327E-02	0.619	3.003E-02	0.461	5.367E-02
0.75	1000	1.178	4.690	2.371E-04	0.933	5.931E-03	0.460	2.378E-02	0.298	5.369E-02	0.215	9.594E-02
	1500	1.767	7.035	2.298E-04	1.403	5.889E-03	0.694	2.369E-02	0.455	5.356E-02	0.333	9.577E-02
	2000	2.356	9.381	2.23E-04	1.872	5.848E-03	0.929	2.361E-02	0.611	5.344E-02	0.450	9.560E-02
0.50	1000	0.785	4.688	5.334E-04	0.922	1.335E-02	0.437	5.354E-02	0.265	1.211E-01	0.170	2.166E-01
	1500	1.178	7.033	5.170E-04	1.391	1.325E-02	0.672	5.336E-02	0.421	1.208E-01	0.287	2.163E-01
	2000	1.571	9.379	5.027E-04	1.860	1.316E-02	0.906	5.317E-02	0.577	1.205E-01	0.404	2.159E-01

It can be inspected from Figure 3 that more noticeable deviations in TPs from the laminar CHF profile (Equation 15) arise in the flow scenarios with lower micro-pipe diameter and heat flux and higher Reynolds number. This determination can be explained by the growth of Ψloss values under these flow conditions (Table 2). The higher energy loss outputs, with lower d and higher Re, count on the enhanced wall velocity gradients (Figure 2) and the corresponding stronger friction activity (Table 1). On the other hand, the source of the raised Ψloss with lower q'' can be enlightened with the decay of ΔT data (Table 2) with this manipulation. Since liquid viscosity responses inversely to fluid temperature, lower ΔT due to reduced q'' results in promoted viscosity levels causing as well the motivation of Ψloss values. Besides, the reverse impact of Re on ΔT can be outlined with the direct relation of Reynolds number and mass flow rate in the micro-pipe, where the overall transferred heat through the micro-pipe surface can also be defined with the fundamental formula of $Qs = m\,.\,Cp\Delta T$. Figure 3 additionally puts forward that the wall temperature gradients $(\partial T(r)/\partial r|_{r/R=1})$ rise with lower d and higher Re, in other words in cases with excited Ψloss. Contributing directly the wall temperature gradients with heat transfer rates (Equation 2b) enlightens the parallelism of frictional and thermal activities from the point of attaining motivation with lower d and higher Re. Depending on the above indicated and verified scientific association of TP formation with Ψloss data, to interpret and discuss the deviations in TP profiles with the considered structural (d), flow (Re) and thermal (q'') parameters, the variations in Ψloss under different operating conditions are identified by Equation 16.

$$\Delta\Psi_{loss} = \left(\Psi_{loss}\right)_{q''=1000\ W/m^2} - \left(\Psi_{loss}\right)_{q''=2000\ W/m^2} \qquad (16)$$

Analyses show that as ΔΨloss gets the values of 7.6×10-6 W and 1.9×10-4 W for Re = 100 and 2,000 in the micro-pipe flow of d = 1.00 mm, these data rise to 1.41×10-5 W and 3.4×10-4 W at d = 0.75 mm and to 3.07×10-5 W and 7.0×10-4 W at d = 0.50 mm. These findings indicate that the potential of q '' on Ψloss develops with lower d and higher Re. This determination is quite significant in identifying the scenarios where the TPs are more probable to deviate from the laminar constant heat flux profile. Since ΔT and Ψloss respond inversely to the imposed parameters, they develop completely in a contrary mechanism under the influence of these operational factors. Furthermore, simultaneous inspection of the depended character of Ψloss on the VPs and the affiliation of ΔT with TPs evidently put forward the strong interaction of the momentum and thermal characteristics in micro-pipe flows.

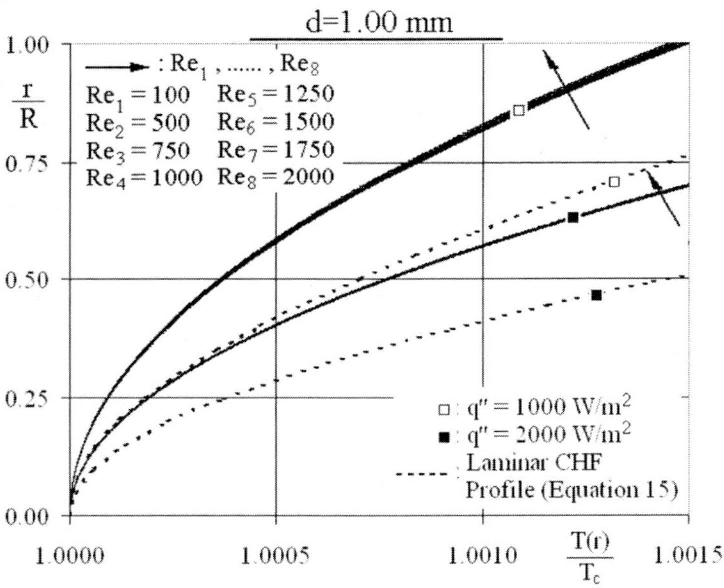

(a)

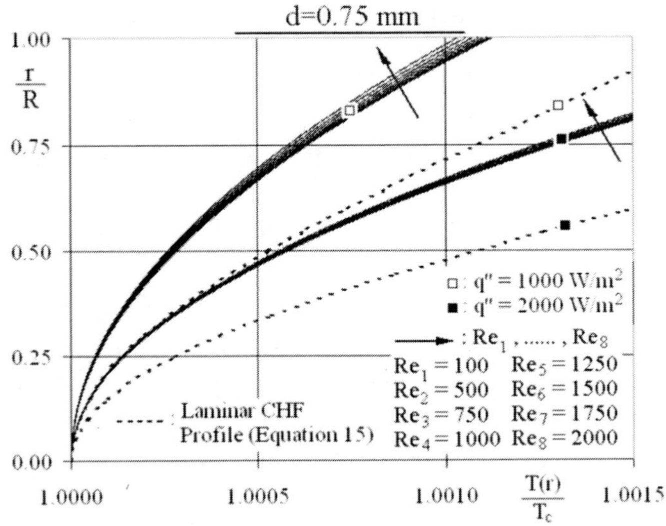

(b)

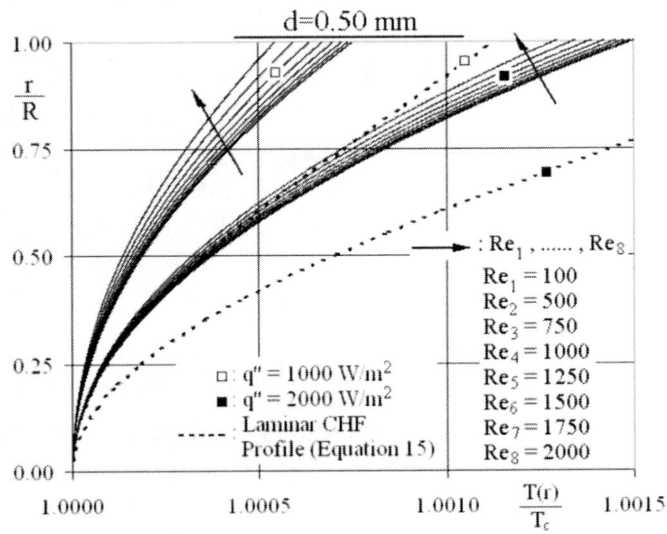

(c)

Figure 3. Variations of radial temperature profiles with Re and q '' for the micro-pipe diameters of (a) d = 1.00 mm, (b) d = 0.75 mm and (c) d = 0.50 mm.

3.1.2. Characteristic parameters

Computational Cf* values, for various Re and d cases, are presented in Figure 4, together with the available literature based data. The roles of Reynolds number and micro-pipe diameter on friction coefficient comprise appreciable importance not only from the point of flow characteristics and momentum transfer, but also due to direct contribution of viscous effects on the local frictional entropy generation (S''' ΔP - Equation 10). It can be seen from Figure 4 that Cf* values augment with both higher Reynolds number and lower micro-pipe diameter. Computations identified the Cf* range for Re = 100 as Cf* = 1.006→1.007 (d = 1.00→0.50 mm), whereas the corresponding ones for Re = 500, 1,000 and 1,500 are Cf* = 1.028→1.038, 1.063→1.075 and 1.088→1.116, respectively. These numbers enlighten the combined contribution of Re and d on the transformation of Cf* values towards the transition onset label of Cf* = 1.1. It can additionally be resolved from the above ranges that as the deviation of (Cf*)d=1.00 mm and (Cf*)d=0.50 mm at Re = 100 is only 0.1%, it becomes 0.97%, 1.13% and 2.57% at Re = 500, 1,000 and 1,500, respectively. These deviations put forward that the impact of micro-pipe diameter on momentum characteristics becomes more significant at high Reynolds numbers. The complete set of presentations and analysis on the characterization of micro-pipe flows with Reynolds number and micro-pipe diameter not only ascertained the decrease of H and Uc/Uo and increase of Cf* and γ values, but also identified the shift process of VPs from fully laminar to the transitional structure. As tendency to transitional activity and promoted friction coefficients account to the augmentation of pressure loss amounts of the flow, the present findings on the individual and combined roles of Reynolds number and micro-pipe diameter are quite remarkable for micro-level industrial applications such as heat exchangers, pumps, electronics cooling, piping arrangements and lubrication systems. On the other hand, Figure 4 as well shows that although the Cf* values of Yu et al. [40] were even below the laminar theory for 100 < Re < 2,000, the experimental friction coefficient data of Vijayalakshmi et al. [37] (ε* = 0.002), Choi et al. [41] (ε* = 0.0003), Kohl et al. [42] (ε* = 0.003–0.013) and Wu and Little [39] exposed strengthening trends for Re > 500, showing harmony with the present numerical evaluations.

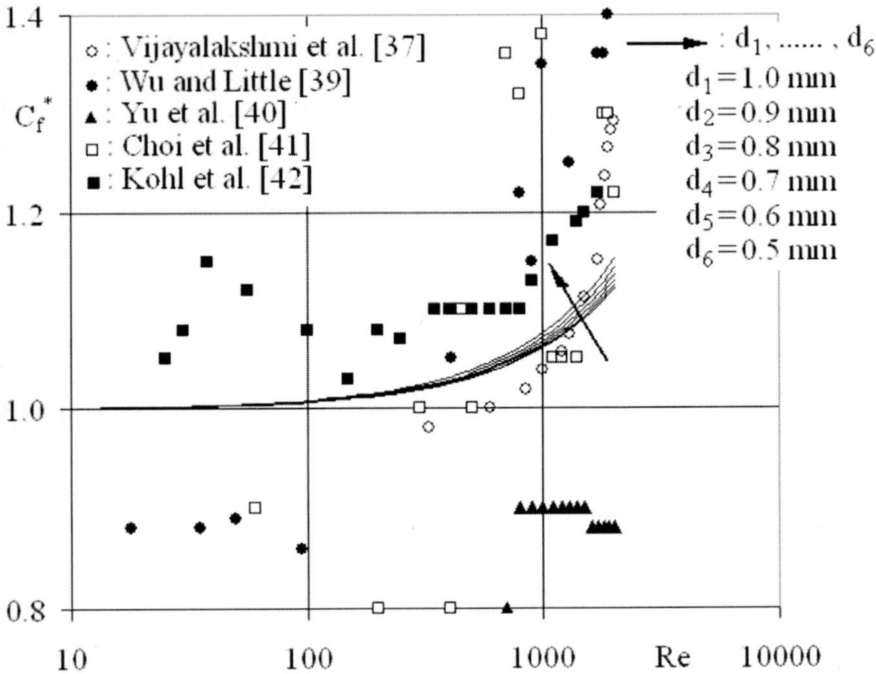

Figure 4. Variation of Cf* with Re and d.

Figure 5 presents the variation of Nusselt numbers for various Reynolds numbers and micro-pipe diameters, where the computational outputs are compared not only with the literature based experimental results but also with the empirical correlations of the related recent work. Analyses shown that Nusselt numbers, for the complete micro-pipe diameter range investigated, attained the almost constant value of Nu = 4.25 at the low Reynolds number range of Re < 100, where this value lays between the characteristic indicators of Nu = 4.36 and Nu = 3.66 for the laminar constant heat flux (CHF) and constant surface temperature (CST) scenarios. The present computational Nu trends in Figure 5 put forward the rise of heat transfer activity in flows with lower micro-pipe diameters and higher Reynolds numbers. Indeed this finding complies with the high wall temperature gradients (Figure 3) at the identical conditions and assists as a double verification marker in the heat transfer vision of this research. Comparing the current numerical results with the experimental reports of Kandlikar et al. [10], Obot [16] and Wu and Little [39] exhibit a close harmony with their upper limit of Nusselt number set for Re ≥ 850. On the other hand, as the analogy of Choi et al. [41] (Equation 17a), developed for the significantly low micro-pipe diameters of d ≤ 0.081 mm, matches with the present outputs only in the narrow Reynolds number band of 700 < Re < 900, Li et al.'s [43] empiric suggestion (Equation 17b) not only displays the same style of the present variations but also its solution set for d = 0.75 mm shows close similarity with the present results of the d = 0.50–0.70 mm cases for Re > 300.

$$Nu=0.0000972*Re^{1.17}*Pr^{0.333} \qquad Nu = 4.1 + \frac{0.14*d/L*Re*Pr}{1+0.05*(d/L*Re*Pr)^{2/3}} \tag{17a-b}$$

To provide a more sensitive designation on the individual role of micro-pipe diameter on Nusselt number, the ratio of Nud=0.50 mm/Nud=1.00 mm is taken into consideration, which proposed the variation rates of 1.065, 1.126, 1.181 and 1.231 at the Reynolds number levels of Re = 500, 1,000, 1,500 and 2,000 respectively. These proportions clearly identify the growing influence of the micro structure on the heat transfer behavior at high Reynolds numbers. Besides, comparing the Nusselt numbers evaluated at the transition Reynolds numbers (Nutra) with the laminar typical value of Nulam = 4.25 points out the augmentation rates of Nutra/Nulam as 1.30 and 1.41 at d = 1.00 mm and d = 0.50 mm, respectively. These values put forward that at the onset of transition, heat transfer activity is more potentially provoked at lower micro-pipe diameter cases. It can additionally be extracted from these proportions that, for the micro-pipe diameter range of d = 0.50–1.00 mm with the fairly low relative roughness of ε* = 0.001, the Nusselt numbers rise by a factor range of 1.30–1.40 at the transition onset.

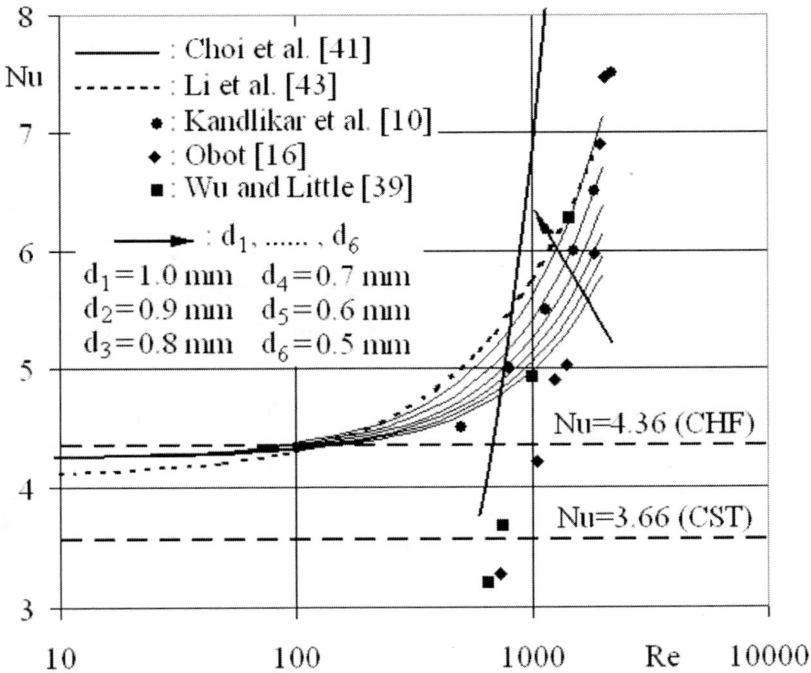

Figure 5. Variation of Nu with Re and d.

3.2. Analyses on Second-Law Characteristics

3.2.1. Radial schemes

Due to the fact that local entropy generation is dependent on the two sub-components (Equation 10b) of thermal ($S'''_{\Delta T}$) and frictional ($S'''_{\Delta P}$) parts, at the initial step of the related discussions the radial distribution characteristics of these individual items are presented.

These discussions will cooperate to enlighten the local influences of pipe diameter (d), Reynolds number (Re) and heat flux (q'') on the mechanism of entropy generation. Figure 6 shows the radial profiles of $S'''_{\Delta T}$, those of $S'''_{\Delta P}$ are displayed in Figure 7.

Figure 6 clearly identifies that as the flow at the pipe centerline (r/R = 0) does not have any contribution on $S'''_{\Delta T}$, thermal based entropy generation values attain the highest rates at the micro-pipe walls (r/R = 1). This outcome can be simultaneously interrelated with the fundamental definition of $S'''_{\Delta T}$ and temperature profiles of Figure 3. Since thermal entropy generation is mainly dependent on temperature gradients in the flow domain (Equations 10a-b), the locations of the least and top gradients also define the locus of the lowest and highest $S'''_{\Delta T}$ values. Moreover, as expanded in detail through Figure 3, the impacts of d, Re and q'' on TP development as well manipulate the thermal entropy generation rates; thus the $S'''_{\Delta T}$ values go along not only the with the TPs but also with the Ψloss and ΔT data (Table 2). To identify the individual and combined roles of d, Re and q'' on $S'''_{\Delta T}$, Equations 18 a and b are evaluated at the pipe wall (r/R = 1) for the considered micro-pipe diameter and Reynolds number ranges. For the highest micro-pipe diameter of d = 1.00 mm, β1 is computed to attain the values of 3.99, 4.03 and 4.14 for Re = 100, 1,000 and 2,000, respectively. As decreasing the pipe diameter to d = 0.75 mm restructures the above figures to 3.99, 4.08 and 4.36, at the lowest diameter (d = 0.50 mm) of the present research, β1 is augmented up to 4, 4.29 and 5.66. Inspecting β2 clarifies the superiority of $S'''_{\Delta T}$ (d = 1.00 → 0.50 mm) with the ratios of 1, 1.13 and 1.78 for Re = 100, 1,000 and 2,000 at q'' = 1,000 W/m2, whereas these proportions fall to 1, 1.06 and 1.30 at q'' = 2,000 W/m2.

$$\beta_1 = \frac{\left(S'''_{\Delta T}\right)_{q''=2000\ \mathrm{W/m^2}}}{\left(S'''_{\Delta T}\right)_{q''=1000\ \mathrm{W/m^2}}} \qquad \beta_2 = \frac{\left(S'''_{\Delta T}\right)_{d=1.00\ \mathrm{mm}}}{\left(S'''_{\Delta T}\right)_{d=0.50\ \mathrm{mm}}} \tag{18a-b}$$

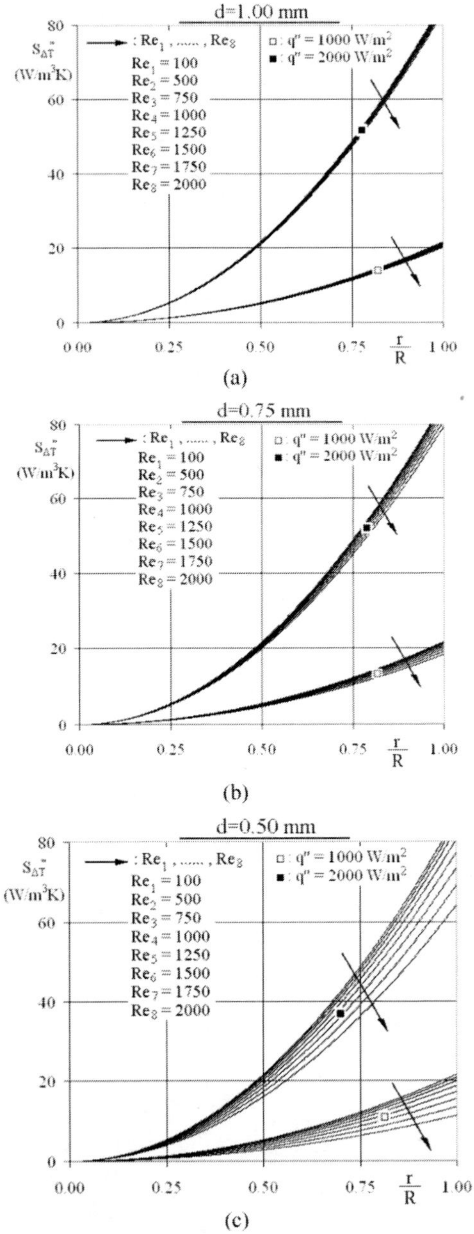

Figure 6. Variation of radial S ''' ΔT distributions with Re and q '' for the micro-pipe diameters of (a) d = 1 mm, (b) d = 0.75 mm and (c) d = 0.50 mm.

From the β1 point of view, it can be said for the low Reynolds numbers (Re = 100) that the encouraging role of q '' on S ''' ΔT is almost fixed in the complete micro-pipe diameter range investigated. However, the influence of heat flux grows with both high Re and low d, especially for Re > 1,000 and d < 0.75 mm, indicating that the intend of flow towards both transitional and micro-flow character are additional motivations to increase the local thermal entropy generation rates. The second indicator (β2) further identifies that the impact of d improves with Re and drops with q '' .

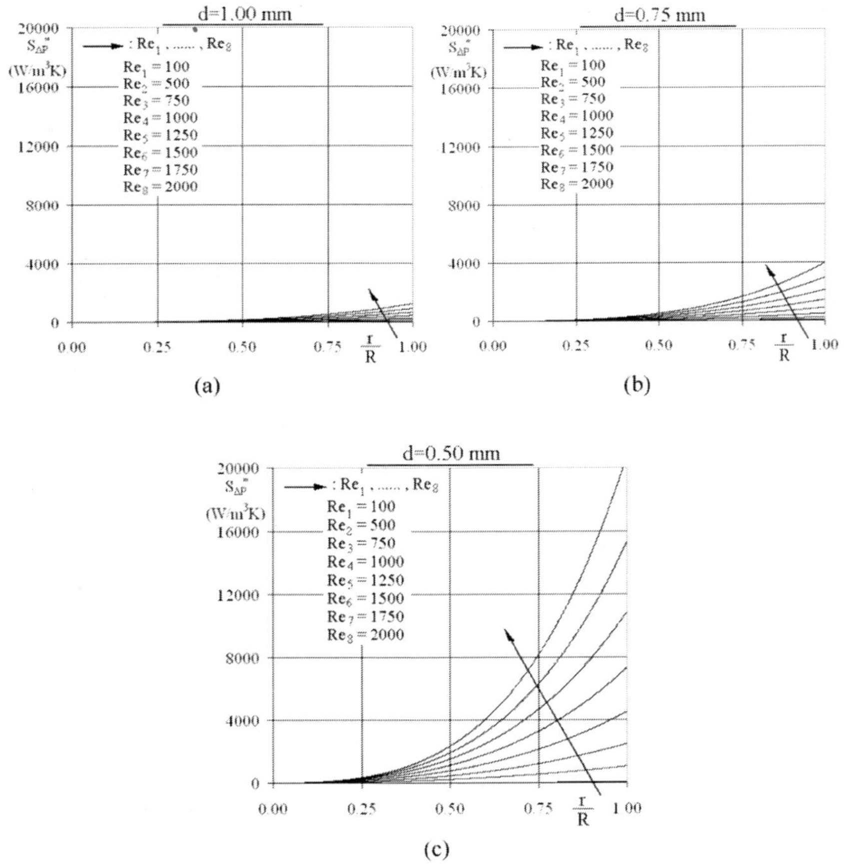

Figure 7. Variation of radial S ''' ΔP distributions with Re for the micro-pipe diameters of (a) d = 1 mm, (b) d = 0.75 mm and (c) d = 0.50 mm.

Figure 7 shows the formation and variation of radial characteristics of frictional entropy generation ($S'''_{\Delta P}$) for various Reynolds numbers and micro-pipe diameters. Since the viscous energy loss values (Ψloss) for different heat flux applications are quite similar (Table 2), only the $S'''_{\Delta P}$ profiles for $q'' = 1{,}000$ W/m2 are presented in Figure 7. It can be seen from the figure that $S'''_{\Delta P}$ values rise towards the pipe walls, where the most significant velocity gradients (Figure 2), shear rates, friction coefficients (Figure 3) and the frictional entropy generation values are located. Due to the fact that frictional entropy generation is essentially defined by the magnitude of velocity gradient, there exist not only scientific but also operational close relation among the Cf* and $S'''_{\Delta P}$ values. As in the variation of friction coefficients with Reynolds number and micro-pipe diameter (Figure 4–Table 1), frictional entropy generation rates are as well computed to enhance with higher Re and lower d. This outcome can be enlightened through the elevated fluid flow-solid domain interaction in cases with narrower flow domains (lower micro-pipe diameter) or increased mass flow rate (higher Reynolds number). Present computational determinations are completely in harmony with the results of Ko [31] for double-sine ducts, on the variations of the thermal and frictional entropy generation rates with Reynolds number.

The individual and combined roles of micro-pipe diameter and Reynolds number can be distinguished more specifically by comparing the wall ($r/R = 1$) frictional entropy generation values of different micro-pipe diameters and Reynolds numbers (Equations 19a-b). Numerical analyses show that α1 has the values of 1.00, 5.62 and 16.90 for Re = 100, 1,000 and 2,000, respectively; besides α2 is computed as 30.99 and 523.42 for the micro-pipe diameters of d = 1.00 mm and 0.50 mm. The grow of α1 with higher Re clearly states that the influence of micro character on frictional entropy generation becomes more impressive at higher Reynolds numbers. Besides, the provoked view of α2 at lower d establishes the accelerated role of Re on $S'''_{\Delta P}$ in the pipes with stronger micro character. Indeed α1 and α2 simultaneously denote the synergy imposed on frictional entropy generation by the concurrent action of Reynolds number and micro-pipe diameter.

$$\alpha_1 = \frac{\left(S'''_{\Delta P}\right)_{d=0.50\ \mathrm{mm}}}{\left(S'''_{\Delta P}\right)_{d=1.00\ \mathrm{mm}}} \qquad \alpha_2 = \frac{\left(S'''_{\Delta P}\right)_{Re=2000}}{\left(S'''_{\Delta P}\right)_{Re=100}} \tag{19a-b}$$

Individual and combined roles of surface heat flux, Reynolds number and micro-pipe diameter on the radial variations of local Bejan number (Be ''') are demonstrated in Figure 8. It can be inspected from the figure that Bejan numbers decrease in the radial direction (r/R = 0→1). This output puts forward that among the both radially increasing two sub-components of entropy generation (S ''' ΔT -Figure 6 and S ''' ΔP -Figure 7), augmentation in the frictional component is more significant. The reduced appearance of Be ''' with Re is an expected evaluation due to the deeply discussed drop of S ''' ΔT and rise of S ''' ΔP with Re (Figure 6 and Figure 7). On the other hand, analyses show that local Bejan numbers increase with both heat flux and micro-pipe diameter. The rise of Be ''' due to q '' can be attributed not only to the q '' dependent grow of S ''' ΔT but also the unvarying nature of S ''' ΔP with q ''. Whereas the role of d on Be ''' identifies the decay of S ''' ΔP, thus viscous losses, and the featured behavior of S ''' ΔT with the depart from the micro-flow character (d→1.00 mm). By comparing the wall (r/R = 1) Bejan number values at different ranks of surface heat flux, Reynolds number and micro-pipe diameter through Equations 20a-b, can provide a deeper insight into the mechanism of Be ''' .

$$\varsigma_1 = \frac{\left(Be'''\right)_{q''=2000\ \mathrm{W/m^2}}}{\left(Be'''\right)_{q''=1000\ \mathrm{W/m^2}}} \qquad \varsigma_2 = \frac{\left(Be'''\right)_{d=1.00\ \mathrm{mm}}}{\left(Be'''\right)_{d=0.50\ \mathrm{mm}}} \tag{20a-b}$$

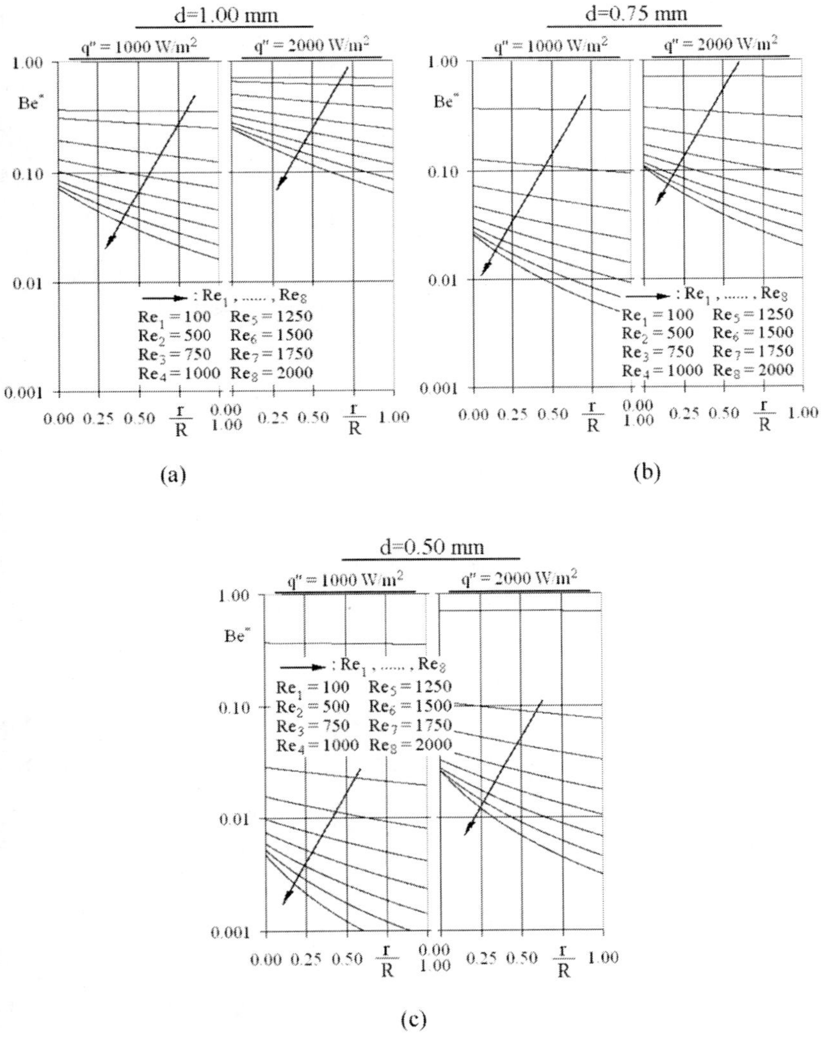

Figure 8. Variation of radial Be ''' distributions with Re and q '' for the micro-pipe diameters of (a) d = 1.00 mm, (b) d = 0.75 mm and (c) d = 0.50 mm.

Computations show that as ζ1 gets the values of 1.94, 3.31 and 3.94 for Re = 100, 1,000 and 2,000, respectively at d = 1.00 mm, they become 1.94, 3.81 and 4.29 at d = 0.75 mm and 1.94, 4.24 and 5.65 at d = 0.50 mm. Since surface heat flux is not influential on frictional entropy generation (Figure 7–Table 2), $\zeta 1$ is mainly structured by the variation of thermal entropy rates. The rising impact of q'' on Be''' with higher Re puts forward that the elevating role of q'' on $S'''\Delta T$ is dominant to the suppressing function of Re on $S'''\Delta T$. Besides, $\zeta 1$ also shows that the variation of Be''' with q'' becomes enhanced at low d, interpreting that the bridling function of q'' on $S'''\Delta T$ is dominant to that of d on $S'''\Delta T$.

The above findings completely enlighten the chief configurative stance of heat flux among the characterizing parameters of Bejan number in micro-pipe flows. It can additionally be extracted from $\zeta 1$ that, the overall scheme of the above defined characteristic actions on Be''' is more significant in the Reynolds number range of Re = 100–1,000; for Re > 1,000 the momentum of the augmenting role decreases. This evaluation puts forward that the exhibited functions is more decisive in laminar flow. On the other hand, $\zeta 2$ figures out the potential of micro-pipe diameter on Bejan number with the descriptive numbers of 1.00, 17.36 and 29.55 for Re = 100, 1,000 and 2,000 at q'' = 1,000 W/m2 and1.00, 13.56 and 20.60 at q'' = 2,000 W/m2. $\zeta 2$ puts forward complementary indicators, not only identifying the more noticeable role of micro-pipe diameter on Bejan number in scenarios with high Reynolds number and low heat flux but also pointing out the comprehensive motivation of micro-pipe diameter on Bejan number in the Reynolds number range of Re = 100–1,000.

3.2.2. Characteristic parameters

Figure 9 displays the cross-sectional total entropy generation (S') in the micro-pipe for various surface heat flux, Reynolds number and diameter cases. The grow of S' with q'' is completely dependent on the rise of $S'''\Delta T$ with q'' (Figure

6), due to the unnoticeable change in frictional entropy generation at different levels of heat flux (Figure 7–Table 2). Besides, the superior S' in flows with lower d can be interrelated not only with the forced S ''' ΔP values at low d but also with the overwhelming increase of S ''' ΔP to the decrease rate of S ''' ΔT in lower micro-pipe diameters. On the other hand, the augmentation of total entropy generation at high Reynolds numbers can be attributed to the dominant progress of frictional entropy generation to the decline of thermal entropy generation towards the upper Reynolds number limit of the present work. The rise of the entropy values with Reynolds number were also reported by Lin and Lee [25] for heat exchangers, by Rakshit and Balaji [28] for a finned channel flow and by Zimparov [29] for a constant wall temperature duct flow.

To identify the individual and integrated influences of q '' , Re and d on S' , the following parameters of λ1 and λ2 are defined (Equations 21 a and b). Calculations resulted in the λ1 values of 2.07, 1.02 and 1.00 for for Re = 100, 1,000 and 2,000, respectively at d = 0.50 mm and 2.07, 1.08 and 1.02 at d = 0.75 mm and 2.07, 1.24 and 1.06 at d = 1.00 mm. Whereas those of λ2 come out to be, for the identical Reynolds numbers, as 4.00, 0.27 and 0.25 at q '' = 1,000 W/m2 and 4.00, 0.33 and 0.27 at q '' = 2,000 W/m2. It can be identified from λ1 that the impact of q '' on S' is more sensible in scenarios with low Re and high d. This output can be explained by the simultaneous presence of unvarying S ''' ΔP with q '' and stimulated S ''' ΔT with low Re and high d.

$$\lambda_1 = \frac{\left(S'\right)_{q''=2000\ \mathrm{W/m^2}}}{\left(S'\right)_{q''=1000\ \mathrm{W/m^2}}} \qquad \lambda_2 = \frac{\left(S'\right)_{d=1.00\ \mathrm{mm}}}{\left(S'\right)_{d=0.50\ \mathrm{mm}}} \tag{21a-b}$$

Furthermore, λ2 additionally clarifies the notable contribution of d on S' at high q '' with low Re. Moreover, the almost heat flux independent character of the micro-pipe diameter effect on cross-sectional total entropy generation (λ2 = 0.25–0.27) at high Reynolds numbers can be confirmed by not only with the predominant $S'''\ \Delta P$ composition in S' but also through the decrease of Be''' with high Re (Figure 8). Since entropy generation is a fundamental scientific indicator of the amount of thermodynamic irreversibilities, the present foresights on the simultaneous actions and roles of Reynolds number, pipe diameter and surface heat flux on the second-law characteristics of micro-pipe flows can be considered as guiding tools in handling the second-law efficiency aspects of micro-level power generation systems or turbomachines, such as gas turbines, compressors, turbines and nozzles.

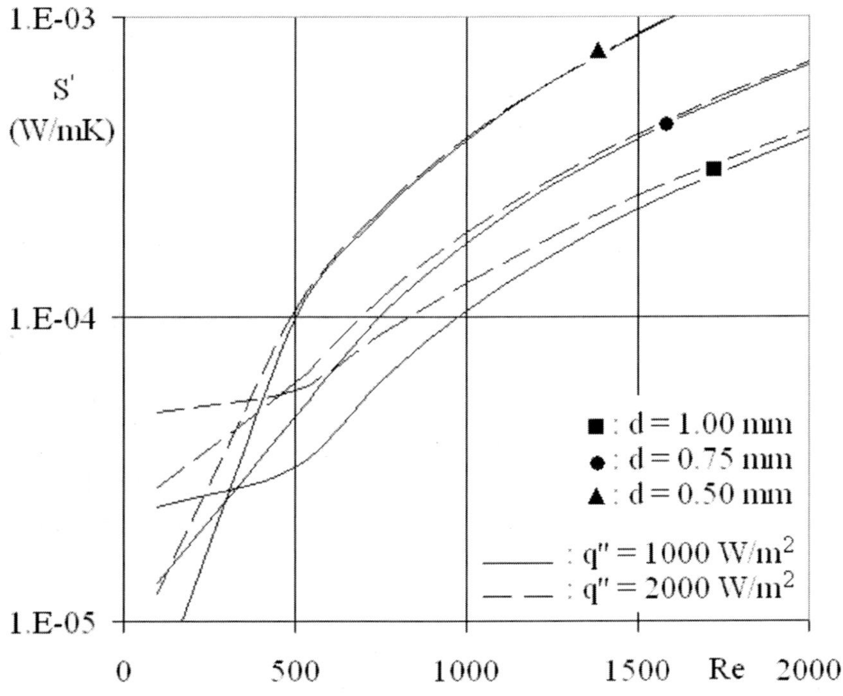

Figure 9. Variation of S' with Re, d and q''.

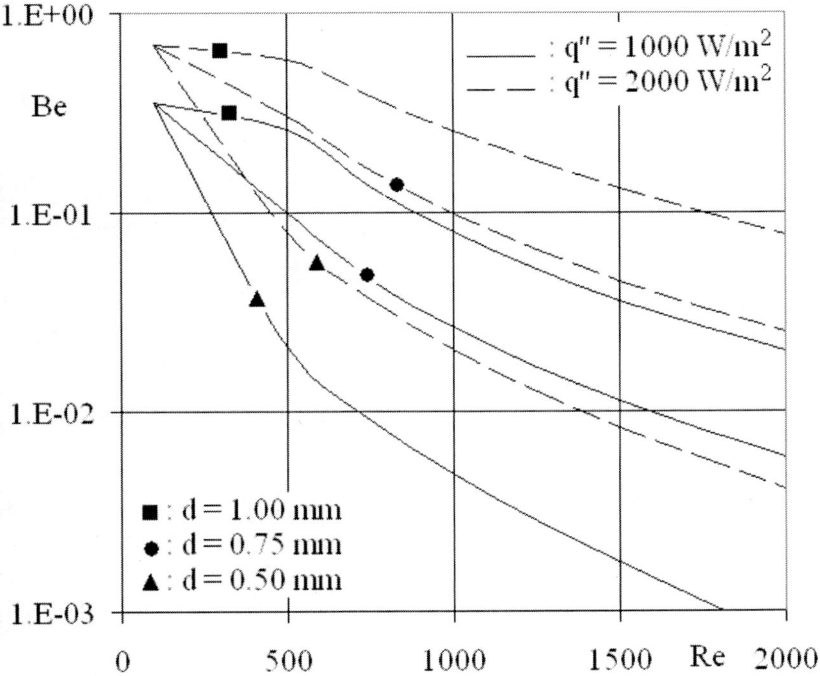

Figure 10. Variation of Be with Re, d and q ''.

Variation of cross-sectional average Bejan numbers with surface heat flux, Reynolds number and micro-pipe diameter is given in Figure 10. As expected from the findings and discussions through Figure 8, Bejan number decreases with high Reynolds number and low micro-pipe diameter and heat flux. To provide a deeper inspection into the second-law mechanism of Be, based on flow (Re), thermal (q '') and structural (d) components, Equations 22a-b are defined in terms of the employed scenarios.

$$\pi_1 = \frac{(Be)_{q''=2000\ W/m^2}}{(Be)_{q''=1000\ W/m^2}} \qquad \pi_2 = \frac{(Be)_{d=1.00\ mm}}{(Be)_{d=0.50\ mm}} \qquad (22a\text{-}b)$$

π1 attains the almost identical values of ζ1 (Be ''') as 1.93, 3.25 and 3.90 for Re = 100, 1,000 and 2,000, respectively at d = 1.00 mm, and 1.93, 3.78 and 4.28 at d = 0.75 mm and 1.93, 4.23 and 5.65 at d = 0.50 mm. The similarity of π1 and ζ1 exposes that the impact of q '' on the local value of Be ''' at r/R = 1 and on the complete Be ''' variation for r/R = 0→1 is mostly matching. On the other hand, π2 data differ from those of ζ2, especially at the high Reynolds number range of Re = 1,000–2,000. At q '' = 1,000 W/m2, π2 are evaluated as 1.00, 16.77 and 28.05 for Re = 100, 1,000 and 2,000 and become 1.00, 12.86 and 19.35 at q '' = 2,000 W/m2. The slightly lower (<6%) view of π2 than ζ2 puts forward that the effectiveness of the extreme centerline Be ''' values (Figure 8) on Be becomes more recognizable in flows with low micro-pipe diameter. Indeed this record is quite significant to designate the flow and thermal fundamentals and thermodynamic interactivity of the radial variations of S ''' ΔT , S ''' ΔP and Be ''' , and also the cross-sectional average Bejan number of Be as well.

4. CONCLUSIONS

Temperature dependent property continuity, momentum and energy equations are simultaneously solved to computationally investigate the flow, heat transfer and second-law characteristics of laminar-transitional micro-pipe flows. The combined effects of pipe diameter, Reynolds number and wall heat flux are presented and discussed in detail.

Analyses on flow characteristics resulted in the following main findings: Friction coefficients increase with higher Reynolds number and/or lower micro-pipe diameter, with comprehensive velocity profile deviations from the laminar profile. As the center velocity values decrease, due to the rise of wall velocity gradients, non-dimensional friction coefficients transform towards the transition with the interpreting figures of Cf* = 1.006→1.007 (d = 1.00→0.50 mm) at Re = 100, Cf* = 1.028→1.038 at Re = 500, Cf* = 1.063→1.075 at Re = 1,000 and Cf* = 1.088→1.116 at Re = 1,500. Towards the transition regime, shape factor data drop below the laminar regime indicator of H = 3.36 and the intermittency elevates. As the combined influence of micro-structure and surface roughness cause earlier transition at the descriptive Reynolds numbers of Retra = 1,656, 1,607, 1,491, 1,341 and 1,272 at d = 1.00, 0.90, 0.75, 0.60 and 0.50 mm respectively, the boundary layer parameters appear in the ranges of H = 3.133→3.159 and γ = 0.137→0.121 (d = 1.00→0.50 mm).

Analyses on heat transfer characteristics identified the following primary issues: The wall temperature gradients are evaluated to rise with lower micro-pipe diameter and higher Reynolds number; besides energy loss come out to be notably excited in these scenarios . Energy loss values additionally rise at lower heat flux applications; moreover the impact of surface heat flux on energy loss is determined to grow at higher Reynolds number and lower micro-pipe diameter. Due to the direct relation of Reynolds number and mass flow rate, Reynolds number inversely motivates the temperature rise of water in the streamwise direction. Heat transfer rates are determined to increase in flows with lower micro-pipe diameters and higher Reynolds numbers. The growing impact of the micro-structure on the heat transfer activity at high Reynolds numbers is revealed by the Nud=0.50 mm/Nud=1.00 mm ratios of 1.065, 1.126, 1.181 and 1.231 at Re = 500, 1,000, 1,500 and 2,000 respectively. Moreover, at the transition onset, Nusselt numbers are computed to augment above the typical laminar value (Nulam = 4.25) by 30→40% (d = 1.00→0.50 mm), pointing out the rising strength of micro-flow affects on the heat transfer mechanism at lower micro-pipe diameters.

Analyses on second-law characteristics pointed out the following essential items: It is determined that, at low Reynolds numbers (Re =

100), the augmenting capacity of heat flux on local thermal entropy generation is almost fixed in the complete micro-pipe diameter range of d = 0.50–1.00 mm; however the influence grows with both high Reynolds number and low micro-pipe diameter. Due to the intensity of the velocity gradients, local frictional entropy generation rises towards the pipe walls. Computations exhibited a synergy imposed on local frictional entropy generation with the combined action of higher Reynolds number and lower micro-pipe diameter. The promoted cross-sectional total entropy generation values in lower micro-pipe diameter flows can be attributed not only to the forced local frictional entropy generation with lower diameters but also to the overwhelming increase of local frictional entropy generation to the decrease rate of local thermal entropy generation in lower micro-pipe diameters. In scenarios with high Reynolds numbers and low micro-pipe diameters and heat flux levels, Bejan numbers are monitored to noticeably decrease. In flows with low micro-pipe diameter, local Bejan numbers, around the centerline vicinity, are comprehensively decisive on the cross-sectional average Bejan numbers, which is relatively important from the point of thermodynamic interactivity of the radial variations of local thermal and frictional entropy generation rates and local and cross-sectional average Bejan numbers.

ACKNOWLEDGMENTS

This work is supported by the Research Fund of the University of Uludag through project number: M(U)-2009/35.

REFERENCES

1. Kolb, G.; Schurer, J.; Tiemann, D. Fuel processing in integrated micro-structured heat-exchanger reactors. J. Power Source. 2007, 171, 198–204.
2. Fan, M.Q.; Sun, L.X.; Xu, F. Feasibility study of hydrogen generation from the milled al-based materials for micro fuel cell applications. Energy Fuels 2009, 23, 4562–4566.
3. Dau, V.T.; Dinh, T.X.; Tanaka, K. Design and simulation of a valveless micro pump. J. Adv. Mech. Des. Syst. Manuf. 2009, 3, 69–75.
4. Riccio, G.; Chiaramonti, D. Design and simulation of a small polygeneration plant cofiring biomass and natural gas in a dual combustion micro gas turbine (BIO_MGT). Biomass Bioenerg. 2009, 33, 1520–1531.
5. Guo, Z.Y.; Li, Z.X. Size effect on microscale single-phase flow and heat transfer. Int. J. Heat Mass Transfer 2003, 46, 149–159.
6. Celata, G.P.; Lorenzini, M.; Morini, G.L.; Zummo, G. Friction factor in micropipe gas flow under laminar, transition and turbulent flow regime. Int. J. Heat Fluid Flow 2009, 30, 814–822.
7. Ogedengbe, E.O.B.; Naterer, G.F.; Rosen, M.A. Slip flow irreversibility of dissipative kinetic and internal energy exchange in microchannels. J. Micromech. Microeng. 2006, 16, 2167–2176.
8. Engin, T.; Dogruer, U.; Evrensel, C.; Heavin, S.; Gordaninejad, F. Effect of wall roughness on laminar flow of bingham plastic fluids through microtubes. ASME J. Fluids Eng. 2004, 126, 880–883.

9. Vicente, P.G.; Garcia, A.; Viedma, A. Experimental study of mixed convection and pressure drop in helically dimpled tubes for laminar and transition flow. Int. J. Heat Mass Transfer 2002, 45, 5091–5105.

10. Kandlikar, S.G.; Joshi, S.; Tian, S. Effect of surface roughness on heat transfer and fluid flow characteristics at low reynolds numbers in small diameter tubes. Heat Transfer Eng. 2003, 24, 4–16.

11. Wen, M.Y.; Jang, K.J.; Yang, C.C. Augmented Heat transfer and pressure drop of strip-type inserts in the small tubes. Heat Mass Transfer 2003, 40, 133–141.

12. Koo, J.; Kleinstreuer, C. Viscous dissipation effects in microtubes and microchannels. Int. J. Heat Mass Transfer 2004, 47, 3159–3169.

13. Celata, G.P.; Cumo, M.; McPhail, S.; Zummo, G. Characterization of fluid dynamic behaviour and channel wall effects in microtube. Int. J. Heat Fluid Flow 2006, 27, 135–143.

14. Celata, G.P.; Morini, G.L.; Marconi, V.; McPhail, S.J.; Zummo, G. Using viscous heating to determine the friction factor in microchannels – an experimental validation. Exp. Therm. Fluid Sci. 2006, 30, 725–731.

15. Morini, G.L. Viscous heating in liquid flows in micro-channels. Int. J. Heat Mass Transfer 2005, 48, 3637–3647.

16. Obot, N.T. Toward a better understanding of friction and heat/mass transfer in microchannels – a literature review. Microscale Thermophys. Eng. 2002, 6, 155–173.

17. Wang, H.; Wang, Y.; Zhang, J. Influence of ribbon structure rough wall on the microscale poiseuille flow. ASME J. Fluids Eng. 2005, 127, 1140–1145.

18. Sheikh, A.K.; Zubair, S.M.; Younas, M.; Budair, M. O. Statistical aspects of fouling processes. Proc. Inst. Mech. Eng. E 2001, 215, 331–354.

19. Sahin, A.Z.; Zubair, S.M.; Al-Garni, A.Z.; Kahraman, R. Effect of fouling on operational cost in pipe flow due to entropy generation. Energy Convers. Manag. 2000, 41, 1485–1496.

20. Cao, B.Y.; Chen, M.; Guo, Z. Y. Effect of surface roughness on gas flow in microchannels by molecular dynamics simulation. Int. J. Eng. Sci. 2006, 44, 927–937.

21. Ozalp, A.A. A Computational study to predict the combined effects of surface roughness and heat flux conditions on converging-nozzle flows. Trans. Can. Soc. Mech. Eng. 2006, 29, 67–80.

22. Ozalp, A.A. Numerical analysis of choked converging nozzle flows with surface roughness and heat flux conditions. Sadhana-Acad. Proc. Eng. 2006, 31, 31–46.

23. Kotas, T.J.; Mayhew, Y.R.; Raichura, R.C. Nomenclature for exergy analysis. Proc. Inst. Mech. Eng. A 1995, 209, 275–280.

24. Zimparov, V. Energy conservation through heat transfer enhancement techniques. Int. J. Energy Res. 2002, 26, 675–696.

25. Lin, W.W.; Lee, D.J. Second-law analysis on wavy plate fin-and-tube heat exchangers. ASME J. Heat Transfer 1998, 120, 797–801.

26. Ratts, E.B.; Raut, A.G. Entropy generation minimization of fully developed internal flow with constant heat flux. ASME J. Heat Transfer 2004, 126, 656–659.

27. Sahin, A.Z. Second-law analysis of laminar viscous flow through a duct subjected to constant wall temperature. ASME J. Heat Transfer 1998, 120, 77–83.

28. Rakshit, D.; Balaji, C. Thermodynamic optimization of conjugate convection from a finned channel using genetic algorithms. Heat Mass Transfer 2005, 41, 535–544.

29. Zimparov, V. Extended performance evaluation criteria for enhanced heat transfer surfaces: heat transfer through ducts with constant wall temperature. Int. J. Heat Mass Transfer 2000, 43, 3137–3155.

30. Richardson, D.H.; Sekulic, D.P.; Campo, A. Low reynolds number flow inside straight micro channels with irregular cross sections. Heat Mass Transfer 2000, 36, 187–193.

31. Ko, T.H. Numerical analysis of entropy generation and optimal reynolds number for developing laminar forced convection in double-sine ducts with various aspect ratios. Int. J. Heat Mass Transfer 2006, 49, 718–726.

32. Ko, T.H. Numerical investigation of laminar forced convection and entropy generation in a helical coil with constant wall heat flux. Numer. Heat Transfer PT A 2006, 49, 257–278.

33. Ozalp, A.A. Roughness induced forced convective laminar-transitional micropipe flow: energy and exergy analysis. Heat Mass Transfer 2008, 45, 31–46.

34. Ozalp, A.A. Entropy analysis of laminar-forced convection in a pipe with wall roughness. Int. J. Exergy 2009, 6, 249–275.

35. Ozalp, A.A. 1st and 2nd Law characteristics in a micropipe: integrated effects of surface roughness, heat flux and reynolds number. Heat Transfer Eng. 2009, 30, 973–987.

36. Incropera, F.P.; DeWitt, D.P. Fundamentals of Heat and Mass Transfer; Wiley: New York, NY, USA, 2001.

37. Vijayalakshmi, K.; Anoop, K.B.; Patel, H.E. Effects of compressibility and transition to turbulence on flow through

microchannels. Int. J. Heat Mass Transfer 2009, 52, 2196–2204.

38. White, F.M. Fluid Mechanics; McGraw-Hill: Singapore, 1988.

39. Wu, P.; Little, W.A. Measurement of friction factors for the flow of gases in very fine channels used for microminiature joule–thompson refrigerators. Cryogenics 1983, 23, 273–277.

40. Yu, D.; Warrington, R.; Baron, R.; Ameel, T. An experimental and theoretical investigation of fluid flow and heat transfer in microtubes. In Proceedings of the 4th ASME/JSME Thermal Engineering Conference, Maui, HI, USA, 19th–24th March 1995; ASME: New York, NY, USA; Volume 1, pp. 523–530.

41. Choi, S.B.; Barron, R.F.; Warrington, R.O. Fluid flow and heat transfer in microtubes. Micromech. Sensor. Actuator. Syst. 1991, 32, 123–134.

42. Kohl, M.J.; Abdel-Khalik, S.I.; Jeter, S.M.; Sadowski, D.L. An experimental investigation of microchannel flow with internal pressure measurements. Int. J. Heat Mass Transfer 2005, 48, 1518–1533.

43. Li, J.; Peterson, G. P.; Cheng, P. Three-dimensional analysis of heat transfer in a micro-heat sink with single phase flow. Int. J. Heat Mass Transfer 2004, 47, 4215–4231.

7
CHAPTER

ENHANCED CONVECTIVE HEAT TRANSFER USING GRAPHENE DISPERSED NANOFLUIDS

Tessy Theres Baby and Sundara Ramaprabhu*

Department of Physics, Alternative Energy and Nanotechnology Laboratory (AENL), Nano Functional Materials Technology Centre (NFMTC), Indian Institute of Technology Madras

ABSTRACT

Nanofluids are having wide area of application in electronic and cooling industry. In the present work, hydrogen exfoliated graphene (HEG) dispersed deionized (DI) water, and ethylene glycol (EG) based nanofluids were developed. Further, thermal conductivity and heat transfer properties of these nanofluids were systematically investigated. HEG was synthesized by exfoliating graphite oxide in H_2 atmosphere at 200°C. The nanofluids were prepared by dispersing functionalized HEG (f-HEG) in DI water and EG without the use of any surfactant. HEG and f-HEG were characterized by powder X-ray diffractometry, electron microscopy, Raman and FTIR spectroscopy.

Thermal and electrical conductivities of f-HEG dispersed DI water and EG based nanofluids were measured for different volume fractions and at different temperatures. A 0.05% volume fraction of f-HEG dispersed DI water based nanofluid shows an enhancement in thermal conductivity of about 16% at 25°C and 75% at 50°C. The enhancement in Nusselts number for these nanofluids is more than that of thermal conductivity.

INTRODUCTION

Most industries use conventional fluids like deionized (DI) water, ethylene glycol (EG), transformer oil, etc., as heat transfer fluids. The efficiency of the heat transfer fluid determines the productivity and lifetime of the equipments, electronic circuits, machines, etc. The efficiency of the heat transfer fluids can be increased by enhancing the thermal conductivity and heat transfer properties. Conventional fluids have low thermal conductivity compared to solid counter parts. Therefore, solid particles with high thermal conductivity are generally added to these fluids to enhance their thermal conductivity. However, the addition of macro- and micro-sized particles can create problems like agglomeration and sedimentation. To avoid these problems Choi, Eastman, and co-workers [1, 2] introduced a new type of fluid called nanofluid wherein nanomaterials are dispersed in base fluids like water or EG. Subsequently many research groups have worked on the thermal conductivity and heat transfer mechanism of different nanomaterials dispersed nanofluids. Several groups have shown enhancement in thermal conductivity with Al_2O_3 and CuO nanoparticles dispersed water and EG based nanofluids [3–5]. The enhancement in thermal conductivity depends on several parameters like, size and shape of the nanomaterials, pH of the base fluid, temperature of the fluid, presence of additives, volume fraction of the nanomaterials, etc.

Similar to thermal conductivity, heat transfer mechanism also plays a crucial role in nanofluids. The use of nanofluids having good heat transfer properties reduces the size of the entire unit thereby increases the efficiency of the unit. Hence, it is necessary to determine the heat transfer performance of various nanofluids under dynamic flow conditions apart from steady state thermal conductivity measurements. The heat transfer measurements have been carried out for different flow conditions, laminar flow, and turbulent flow by several groups. Yang et al. [6] studied the heat transfer performance of several nanofluids under laminar conditions in a horizontal tube heat exchanger. Heris et al. [7] found heat transfer enhancement as high as 40% with Al_2O_3 particles. However, there is not much work on the heat transfer mechanism of carbon based nanofluids except a few on carbon nanotubes (CNTs) [8].

Recently, the two-dimensional one carbon atom thick graphene was found to exhibit high crystal quality and ballistic electron transport at room temperature. Theoretical study of thermal conductivity on graphene suggests that it is having unusual thermal conductivity [9, 10]. Following this, Balandin et al. [11] measured experimentally the thermal conductivity of about 5300 W/mK for a single layer graphene from the dependence of the Raman G peak frequency on the excitation laser power. The thermal conductivity of single layer graphene is higher than that of CNTs.

To our knowledge, there is no work on the heat transfer properties of graphene based nanofluids. In the present work, we have synthesized graphene dispersed nanofluids and studied its thermal conductivity and heat transfer properties. The nanofluids were prepared by taking DI water and EG as base fluids.

EXPERIMENTAL METHODS

Materials

Graphite (99.99%, 45 μm) was purchased from Bay Carbon, Inc, USA. All other reagents like sulfuric acid, nitric acid, sodium nitrate, potassium permanganate, hydrogen peroxide, and ethylene glycol were analytical grade. DI water was used throughout the experiment. Graphite oxide (GO) was prepared from graphite using Hummers method [12]. Briefly, 2 g of graphite was treated with 46 ml of sulphuric acid in an ice bath. One gram of sodium nitrate was added to the above solution slowly, followed by the addition of 6 g of potassium permanganate. At room temperature, specific quantity of water was added to the above mixture. After 15 min the suspension was further treated with hydrogen peroxide and was filtered. Finally the filter cake was washed with copious quantity of DI water. At last, the suspension was filtered and dried in vacuum oven at 40°C for 8 h. The dried GO was used for synthesizing hydrogen exfoliated graphene (HEG). Exfoliation of GO was done in hydrogen atmosphere at 200°C as reported previously [13]. Functionalization of HEG was done by treating as synthesized HEG with conc. H_2SO_4:HNO_3 in the ratio 3:1. The acid-HEG mixture was ultrasonicated for about 3 h at room temperature. After 3 h the sample was washed several times with DI water, filtered and dried in vacuum.

Characterization techniques

The samples were characterized with different characterization techniques. Powder X-ray diffraction (XRD) studies were carried out using a PANalytical X'PERT Pro X-ray diffractometer with Nickel-filtered Cu K_α radiation as the X-ray source. The pattern was recorded in the 2θ range of 5° to 90° with a step size of 0.016°. The Raman spectra were obtained with a WITEC alpha 300 Confocal

Raman spectrometer equipped with Nd:YAG laser (532 nm) as the excitation source. Identification and characterization of functional groups were carried out using PerkinElmer FT-IR spectrometer in the range 500-4000 cm^{-1}. Digital photograph has been taken with a Canon Power Shot A590 IS 8 Megapixel camera with 4 × optical zooming. Field emission scanning electron microscopy (FESEM) and transmission electron microscopy (TEM) images were obtained using, FEI QUANTA and JEOL TEM-2010F instruments, respectively. Nanofluid was prepared by dispersing a known amount of f-HEG in the base fluid by ultrasonication (30-45 min). Thermal conductivity of the suspension was measured using KD2 pro thermal property analyzer (Decagon, Canada). The probe sensor used for these measurements were of 6 cm in length and 1.3 mm in diameter. In order to study the temperature effect on thermal conductivity of nanofluid a thermostat was used. Electrical conductivity of the nanofluid was measured using ELICO Ltd CM 183, EC-TDS meter.

The convective heat transfer mechanism was studied using an indigenously fabricated setup. The schematic of the setup is shown in Figure 1. It consists of a flow loop, a heat unit, a cooling part, and a measuring and control unit. The flow loop included a pump with flow controlling valve system, a reservoir and a test section. A straight stainless steel tube with 108 cm length and 23 mm inner diameter was used as the test section. The whole test section was heated by a copper coil linked to an adjustable DC power supply. There was a thick thermal isolating layer surrounding the heater to obtain a constant heat flux condition along the test section. Four T-type thermocouples were mounted on the test section at axial positions in mm of 298 (T1), 521 (T2), 748 (T3), and 858 (T4) from the inlet of the test section to measure the wall temperature distribution, and two further T-type thermocouples were inserted into the flow at the inlet and exit of the test section to measure the bulk temperatures of nanofluids. Cooling part is to cool down the nanofluid coming out from the outlet of test section.

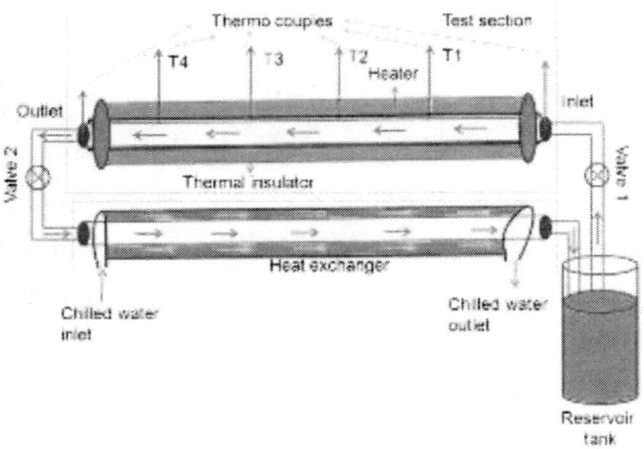

Figure 1. Schematic of the heat transfer set up.

RESULTS AND DISCUSSION

XRD and Raman analysis

The crystallinity of the samples was studied using XRD. Figure 2a shows the X-ray diffractogram of HEG and f-HEG. The characteristic (002) plane in graphite at approximately 26° is shifted to approximately 24° in HEG. This is same in all graphene prepared by different exfoliation techniques [14, 15]. There is not much difference in XRD of f-HEG and HEG except the broadening of the (002) peak. After vigorous acid treatment, the layers might have separated further and that may be the reason for this broadening [16]. The functionalization and defects on the carbon based materials can be sought out by Raman spectroscopy [17]. Figure 2b shows the Raman spectrum of HEG and f-HEG. In HEG, the peak around 1588 cm^{-1} corresponds to the G-band and the peak around 1356 cm^{-1} corresponds to D-band. The D- and G-band represent the sp^3 and sp^2 hybridization of carbon atoms present in the sample, respectively. In the case of f-HEG, G-band as well as D-band shifted to higher wave number side and also broadened with respect to HEG peak positions. G-band has a broad peak centered around 1591 cm^{-1}

and D-band has a peak centered around 1371 cm^{-1}. The ratio of the D-band intensity to G-band intensity in f-HEG is higher than that of HEG. The increase in the relative intensity of the disordered mode can be attributed to the increased number of structural defects and to the sp^3 hybridization of carbon for chemically induced disruption of the hexagonal carbon order after acid treatment. Acid treatment created some functional groups at the edges of the graphene sheets which helped for the proper dispersion of f-HEG on water and EG. The presence of functional groups may be the reason for broadening of the D-band peak.

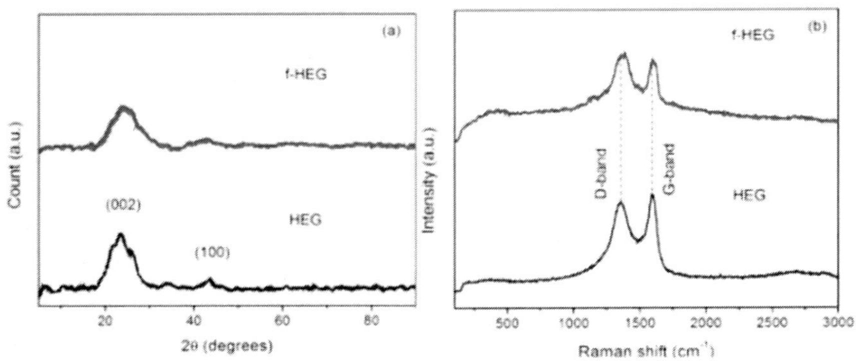

Figure 2. Crystallographic study (a) x-ray diffractogram and (b) raman spectra of heg and f-heg.

FTIR study and digital photograph

The effect of acid treatment and attachment of functional groups were further confirmed with FTIR. Figure 3a shows the FTIR spectra of HEG and f-HEG. During the heat treatment in H_2 atmosphere, most of the oxygen containing functional groups has been removed from HEG. So the functional groups are not dominant in FTIR of HEG. After acid treatment, functional groups are formed at the plane and edges of the sheets. The peaks at around 3442 and 1625 cm^{-1} are due to OH functional groups. A small doublet peak of CH_2 (2922 and 2860 cm^{-1}) and CH at 1365 cm^{-1} are present both in HEG and f-HEG. The peaks at

1720 and 1380 cm^{-1}can be assigned to the C = O and C-O stretching vibrations of COOH. These functional groups help graphene sheets to interact with water molecules and disperse properly. Figure 3b shows the digital photograph of f-HEG dispersed DI water and EG based nanofluid after 2 months of the nanofluid preparation. Even after 2 months of preparation no sedimentation was observed.

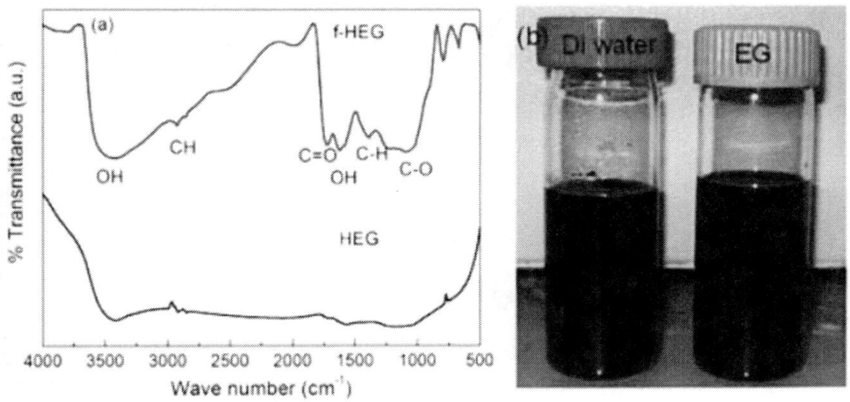

Figure 3. Spectral analysis (a) ftir spectrum of heg and f-heg; (b) digital photograph nanofluid made up of f-heg in di water and eg.

Morphology of graphene sheet

Figure 4a shows the FESEM image of as-synthesized HEG taken by putting a small amount of powder sample on carbon tape. The image shows a large area of transparent graphene sheet with rough and soft wrinkled surface morphology. Transmission electron microscopy is also a powerful technique used extensively to provide definitive identification of graphene materials. The sample preparation was done by depositing a drop of ethanol dissolved HEG on Cu grid. Figure 4b clearly shows the wrinkles on the surface and folding at the edges of HEG sheets.

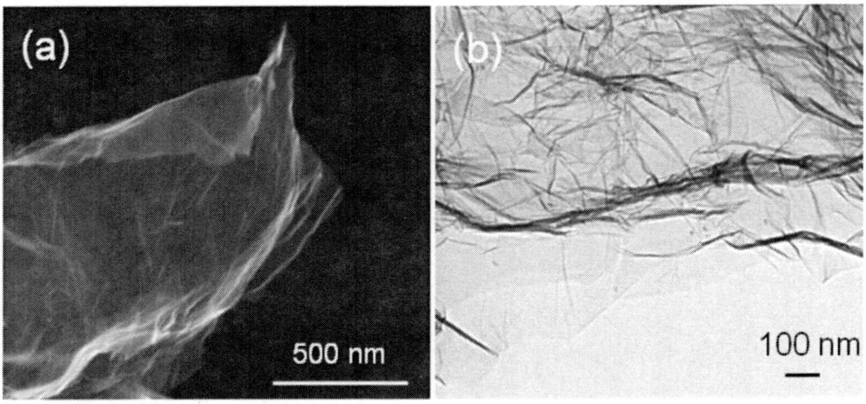

Figure 4. Microscopy study (a) field emission scanning electron microscopy and (b) transmission electron microscopy images of heg.

Thermal conductivity study of graphene nanofluid

The proper functionalization helped to make well-dispersed HEG nanofluid. Figure 5a shows the normalized thermal conductivity (K_n/K_f) of f-HEG dispersed DI water based nanofluid as a function of temperature for different f-HEG volume fractions. All the measurements were carried out for low volume fractions so as to keep the viscosity of the fluid at a minimum level. For DI water based f-HEG nanofluids, the range of volume fractions used was from 0.005 to 0.05%. The percentage enhancement in thermal conductivity was calculated using the relation $((K_n - K_f) \times 100)/K_f$, where '$K_f$' was the thermal conductivity of base fluid and 'K_n' was that of nanofluid. For 0.05% volume fraction, the enhancement in thermal conductivity is about 16% at 25°C and about 75% at 50°C. The enhancement is less than 10% for 0.005% volume fraction. It is clear from the graph that the thermal conductivity increases with increasing temperature and volume fractions. According to Das et al. [4], in nanofluid the main mechanism of thermal conductivity enhancement can be thought as the stochastic motion of the nanoparticles. This Brownian-like motion will be dependent on fluid temperature and so the huge enhancement in thermal conductivity with temperature is quite explicable. At low temperature this motion was less significant giving the characteristics of normal slurries which rapidly changed at elevated temperature bringing more nanoeffect in the conducting behavior of the fluid. The error bar is shown only for low and high volume fractions.

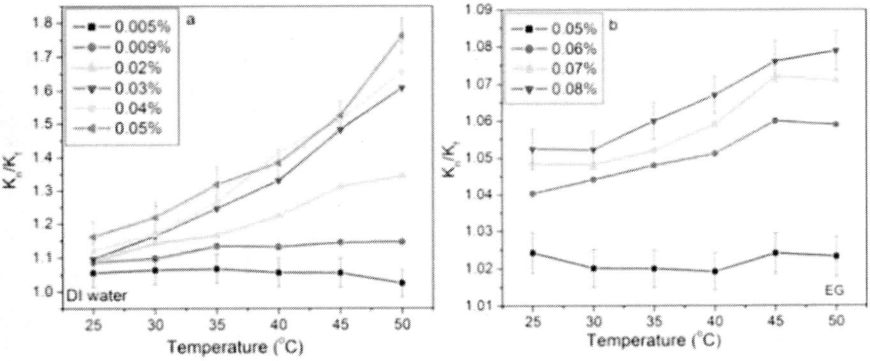

Figure 5. Normalized thermal conductivity of f-heg dispersed. **(A)** DI water and **(B)** EG based nanofluids for different volume fractions and at varying temperatures.

Figure 5b shows the normalized thermal conductivity of f-HEG dispersed EG based nanofluids with varying temperatures and volume fractions. The thermal conductivity of EG based nanofluids did not show much enhancement for low volume fractions. Till around 0.05% volume fraction there was no enhancement in thermal conductivity. Thermal conductivity started increasing from 0.05% volume fraction onwards. For 0.08% the enhancement was about 1% at 25°C and about 5% at 50°C. This low enhancement in thermal conductivity may be due to the high viscosity of EG. Even though the enhancement in thermal conductivity of EG based nanofluids with f-HEG is low, it is slightly higher than that of CNT dispersed EG [18].

The conventional theoretical models on thermal conductivity of nanoparticles suspended fluids does not consider particle size, shape, the distribution, and the motion of dispersed particles, while only thermal conductivities of base fluid and particles, and volume fraction of article. Maxwell was the man who first investigated the thermal conductivity of liquid suspensions analytically [19]. Later, Hamilton and Crossser modified Maxwell's model by taking in to consideration of geometry of particles [20]. In 1987 Hasselman and Johnson [21] modified Maxwell's model by including the interfacial thermal resistance (Kapitza resistance, R_{bd}). The resulting theoretical prediction for the effective thermal conductivity (K) enhancement of the particle-in-liquid colloidal suspensions is given by,

$$K = \frac{K_p(1-2\alpha) + 2K_f + 2[K_p(1-\alpha) - K_f]\varphi}{K_p(1-2\alpha) + 2K_f - [K_p(1-\alpha) - K_f]\varphi} K_f \tag{1}$$

where $\alpha = 2\, R_{bd}K_f\,/d$, d is the average particle diameter, R_{bd} is the interfacial thermal resistance, K_f and K_p are the thermal conductivity of base fluid and particles, respectively. This is also called Maxwell-Garnett type effective medium approximation (MG-EMA). In the absence of thermal boundary resistance ($R_{bd} = 0$), the above equation reduces to Maxwell's model. The results are shown in Figure 6a,b for DI water and EG, respectively. The thermal conductivity is correlated with lower and upper bounds of R_{bd}. The lower and upper bound MG-EMA correlation is almost matching for the DI water based nanofluid at 25°C. When the temperature increases the thermal conductivity is going away from the correlated values. But in the case of EG based nanofluids the calculated value is very much less than the correlated value. This suggests that there are other mechanisms contribute to the thermal conductivity of EG based f-HEG dispersed nanofluids.

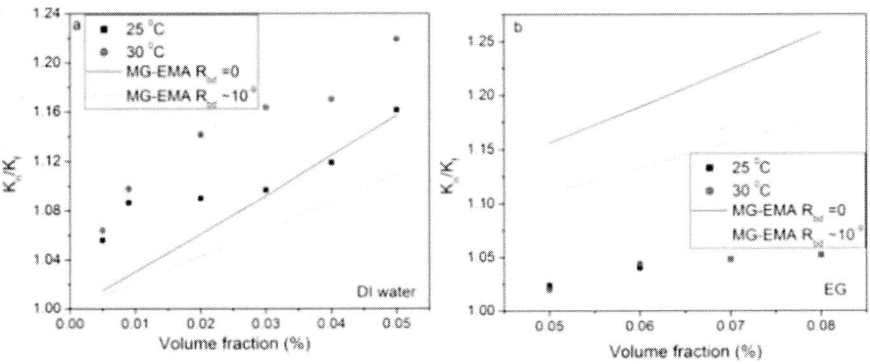

Figure 6. Correlation of experiment with theory. The enhancement of effective thermal conductivity of K_n/K_f as a function of volume fraction correlated for Maxwell-Garnet effective medium approximation (MG-EMA) for **(A)** DI water and **(B)** EG based f-HEG dispersed nanofluids.

Electrical conductivity of f-HEG dispersed nanofluids

Finally electrical conductivity was also measured for some volume fractions of f-HEG dispersed nanofluid. Figure 7a shows the normalized electrical conductivity (σ_n/σ_f) for three different volume fractions at varying temperature in DI water based nanofluid. 'σ_n' represents the electrical conductivity of nanofluid and 'σ_f' that of base fluid. The graph suggests that like thermal conductivity, electrical conductivity also increased with increase in volume fraction and increase in temperature. Similar trend was observed for f-HEG dispersed EG nanofluids also. Figure 7b shows the normalized electrical conductivity of f-HEG dispersed EG based nanofluid for three different volume fractions at varying temperature. The experiments were repeated several times and the error in measurements was less than 4%.

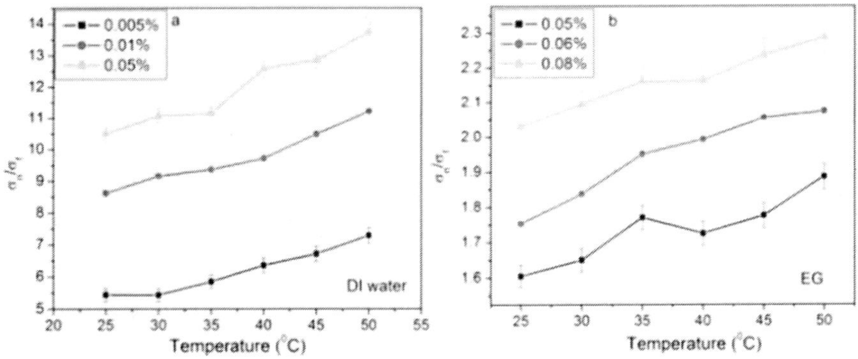

Figure 7. Normalized electrical conductivity of f-heg dispersed (a) di water and (b) eg based nanofluids for different volume fractions and varying temperatures.

Convective heat transfer

The heat transfer coefficient, (h) is a macroscopic parameter describing heat transfer when a fluid is flowing across a solid surface of different temperature. It is not a material property. The convective heat transfer coefficient is defined as

$$h = \frac{q}{(T_s(x) - T_f(x))} \tag{2}$$

where x represents axial distance from the entrance of the test section, q is the heat flux, T_s is the measured wall temperature, and T_f is the fluid temperature decided by the following energy balance:

$$T_f = T_{in} + \frac{E(x)}{Mc_p} \tag{3}$$

where c_p is the heat capacity, M is the mass flow rate, and $E(x)$ is the energy at position x. Equation 3 is based on an assumption of zero heat loss through the insulation layer.

$$E(x) = \frac{(\text{Total energy} \times x)}{\text{Length of the tube}} \tag{4}$$

And mass flow rate can be calculated using the relation,

$$M = uA\rho \tag{5}$$

where u is the velocity of flow, A is the area of cross-section, and ρ is the density of fluid. Reynolds number is defined as $Re = \rho uD/\mu$ and the Prandtl number is defined as $Pr = \nu/\alpha$, where μ is the fluid dynamic viscosity, ν is the fluid kinematic viscosity, and α is the fluid thermal diffusivity.

Validity of the experimental setup with DI water

To check the reliability and accuracy of fabricated experimental setup, systematic measurements were carried out using DI water as the working fluid for different flow rates. The experimental results obtained for different flow rates were correlated with well-known Shah correlation [22] and Dittus-Boelter [23] equation under the constant heat flux boundary condition. The famous Shah correlation is

$$Nu = \begin{cases} 1.953\left(RePr\dfrac{D}{x}\right)^{\frac{1}{3}} & \left(RePr\dfrac{D}{x}\right) \geq 33.3 \\ 4.364 + 0.0722RePr\dfrac{D}{x} & \left(RePr\dfrac{D}{x}\right) < 33.3 \end{cases} \tag{6}$$

where Nu is the Nusselts number. The experimental values were reasonably in good agreement with the Shah equation as shown in Figure 8a. The same was observed for other laminar flow rates also. Reynolds number greater than 10,000 has been correlated with Dittus-Boelter equation given below:

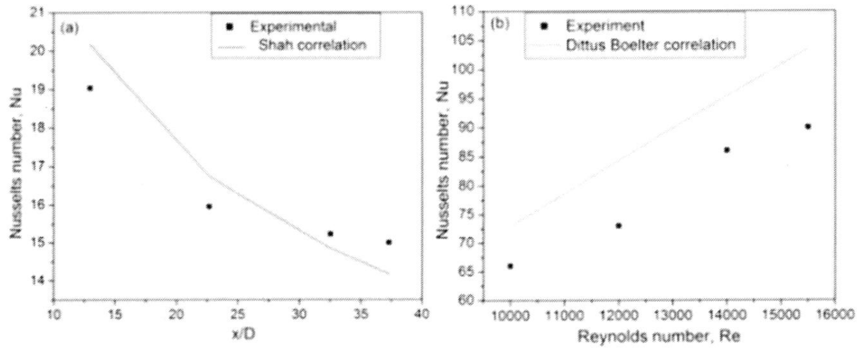

Figure 8. Validity of the experimental heat transfer setup for (a) low (shah correlation) and (b) high (dittus-boelter correlation) flow rates using water.

$$Nu = 0.023Re^{0.8}Pr^{0.4} \tag{7}$$

As shown in Figure 8b, the good coincidence between the experimental results and the calculated values for water reveals that the precision of the experimental system is considerably good. The uncertainty of the experimental system is less than 8%.

Convective heat transfer of graphene nanofluids

Having established confidence in the experimental system, systematic experiments were performed at different flow conditions (Reynolds numbers) for different f-HEG volume fractions under a constant heat flow. From the experiment heat transfer coefficient was calculated and then converts it into corresponding Nusselts number. The Reynolds number is calculated based on the viscosity of the host liquid. Since the calculated Reynolds numbers were greater than 4000, for DI water based nanofluids, the flow was considered to be turbulent. Figure 9a shows the heat transfer measurement of DI water, 0.005 and 0.01% volume fractions f-HEG dispersed DI water for different Reynolds numbers. *X*-axis shows the ratio of axial distance to diameter of the tube (x/D) and Y-axis the corresponding Nusselts number. Black dotted lines, blue solid lines, and red dashed lines are for DI water alone, 0.005% of f-HEG and 0.01% of f-HEG, respectively. Symbols represents Re = 4500 (■),Re = 8700 (●), and Re = 15500 (▲). For better understanding the change in Nusselts number for different Reynolds number is shown in Figure 9b. Similar measurements on EG based nanofluid for different volume fractions and varying Reynolds number are shown in Figure 10. Black dotted lines, blue solid lines, and red dashed lines are for EG alone, 0.005% of f-HEG and 0.01% of f-HEG, respectively. Symbols represents Re = 250 (■),Re = 550 (●), and Re = 1000 (▲). Since the calculated Reynolds numbers were less than 2800, for EG based nanofluids, the flow rates used were laminar.

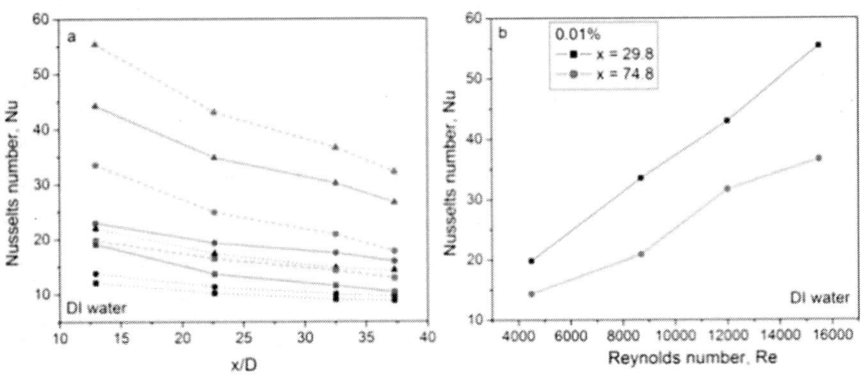

Figure 9. Convective heat transfer study. **(A)** Heat transfer measurement of f-HEG dispersed DI water based nanofluids. Black dotted lines, blue solid lines, and red dashed lines are for DI water alone, 0.005% of f-HEG and 0.01% of f-HEG, respectively. Symbols represents Re = 4500 (*square*), Re = 8700 (*circle*), and Re = 15,500 (*triangle*). **(B)** Measurement of Nusselts number with respect to different Reynolds numbers for DI water based nanofluids, containing 0.01% volume fraction of f-HEG.

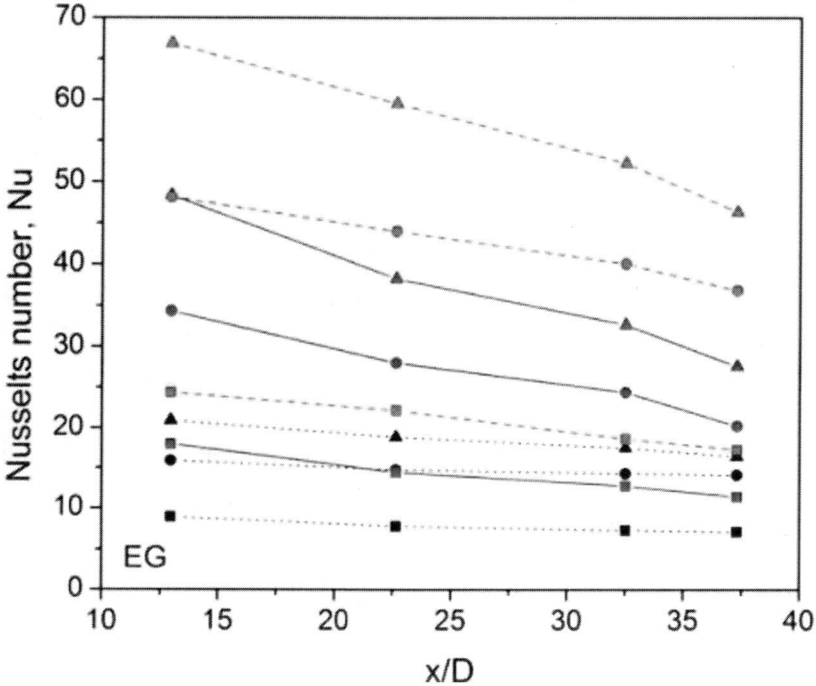

Figure 10. Heat transfer measurement of f-heg dispersed eg based nanofluids. Black dotted lines, blue solid lines, and red dashed lines are for EG alone, 0.005% of f-HEG and 0.01% of f-HEG, respectively. Symbols represents *Re* = 250 (*square*), *Re* = 550 (*circle*), and *Re* = 1000 (*triangle*).

Both the DI water and EG based nanofluids results suggests that the presence of nanomaterials dispersed nanofluids increases the Nusselts number significantly, and the increase is considerably more at high volume fractions and high Reynolds numbers. From Figure 9 it is clear that for a given f-HEG volume fraction, the Nusselts number decreases with axial distance. This is as expected for heat transfer in the entrance region. The percentage enhancement in heat transfer is calculated using the relation $[h_n(x) - h_f(x)] \times 100/h_f(x)$, where $h_f(x)$ and $h_n(x)$ are the heat transfer coefficient for the base fluid and nanofluid at distance *x*, respectively. The enhancement in heat transfer for *Re* = 4500 at the tube entrance is about 64 and 76% for 0.005 and 0.01% volume fractions, respectively. At the outlet, the value decreases to about 21 and 57%, respectively, for 0.005 and 0.01%. When the Reynolds number increases (*Re* = 15,500) the enhancement also increases and it is about 108 for 0.005% and 171 for 0.01% at the entrance. At the end, the values change to about 92 for 0.005% and 141 for 0.01%, respectively.

Similar trend is observed in the case of EG based nanofluid also. Figure 10 shows the variation of Nusselts number for 0.005 and 0.01% f-HEG dispersed EG based nanofluids. From graph it is clear that heat transfer increases with volume fraction. The enhancement in heat transfer for *Re* = 250 at the tube entrance is about 100 and 172% for 0.005 and 0.01%, respectively. At the exit, the value decreases to about 59 and 140%, respectively, for 0.005 and 0.01%. Like water based nanofluids, here also the Nusselts number increases with increase in Reynolds number and it is around 130 and 219% for 0.005 and 0.01% volume fractions, respectively, at the entrance for *Re* = 1000. At the tube exit, the values change to about 69% for 0.005% and 183% for 0.01%. The enhancement in Nusselts number for EG based nanofluids are higher than that of DI water based nanofluids.

Figure 9b shows the effect of the Reynolds number on heat transfer. Figure clearly shows that the Nusselts number increases with increasing Reynolds number. There is a large difference in the Nusselts number at *Re* = 4500 and that at *Re* = 15,500 for DI water based nanofluids. Similar will be the case for EG based nanofluids also (figure not given). This suggests that Reynolds number has a significant effect on the heat transfer mechanism. The enhancement in heat transfer is very drastic compared to the enhancement in thermal conductivity. Another important observation is that even though enhancement in thermal conductivity is very low, enhancement in heat transfer is high for EG based nanofluid.

The reason for decrease in heat transfer from entrance to exit of the tube is due to the variation of thermal boundary layer. In a simple way heat transfer can be written as k/δ with δ the thickness of thermal boundary layer. At the entrance ($x = 0$), the theoretical boundary layer thickness is zero, hence the heat transfer coefficient approaches infinity. The boundary layer increases with axial distance until fully developed after which the boundary layer thickness and hence the convective heat transfer coefficient is constant [8]. Since there is not much enhancement in thermal conductivity, the effect of thickness of thermal boundary may be the reason for this huge enhancement in heat transfer.

Ding et al. [8] also showed that for nanofluids containing 0.5 wt% CNT, the maximum enhancement reaches over 350% at $Re = 800$ and showed that enhancement is a function of the axial distance from the inlet of the test section. Similar observations but with less significant enhancement was observed by Xuan and Li [24] in the turbulent flow regime. Wen and Ding [25] also showed similar features at the entrance region in the laminar flow regime when they investigated heat transfer of aqueous c-alumina nanofluids. They have observed around 47% increase in the convective heat transfer coefficient for 1.6 vol.% nanoparticles loading and $Re = 1600$, which is much greater than that due to the enhancement of thermal conduction (<~10%).

According to the Brownian theory [26], the smaller the sizes of the colloid particles, the faster the particles move, so that energy transport inside the liquid becomes stronger. The clustering or restacking of graphene is very less in solution. Each sheet will be separated out during ultrasonication and was well dispersed which helps for fast heat transfer. Another factor which helps in the enhancement of thermal conductivity as well as heat transfer is surface area of the material. The surface area of hydrogen exfoliated graphene is approximately 450 m^2/g [13]. Other factors which affect the thermal conductivity and heat transfer of nanofluids are size and shape of the nanomaterials, the material (test section) in which nanofluid is flowing, temperature of the test section as well as the surrounding, viscosity of the fluid, etc. Further studies are being carried out for deeper understanding of the mechanism.

CONCLUSION

Graphene was synthesized by hydrogen induced exfoliation of graphite oxide. Further we effectively dispersed f-HEG in the base fluids without any surfactant or additives by using ultrasonication. Systematic characterization and experiments were carried out for the sample preparation as well as the thermal conductivity and heat transfer measurements. The results suggest that there was considerable enhancement in thermal conductivity and heat transfer for f-HEG dispersed fluid compared to its base fluid. The Nusselts number increases with increase in volume fraction and Reynolds number of f-HEG. Similarly, the thermal conductivity of f-HEG increases due to the increase in volume fraction and temperature. Electrical conductivity of f-HEG dispersed base fluids was also showing enhancement compared to the base fluid.

ABBREVIATIONS

DI: deionized

EG: ethylene glycol

FESEM: field emission scanning electron microscopy

GO: graphite oxide

HEG: hydrogen exfoliated grapheme

TEM: transmission electron microscopy

XRD: X-ray diffraction.

ACKNOWLEDGEMENTS

The authors like to thank Indian Institute of Technology Madras, India for giving the financial support for the present work.

AUTHORS' CONTRIBUTIONS

TTB has carried out the experimental part, set up the experimental setup, analysis of the data and manuscript preparation. SR has participated in the design of the experimental setup,, analysis of the data and drafted the manuscript.

REFERENCES

1. Choi SUS, Zhang ZG, Yu W, Lockwood F, Grulke EA: Anomalous thermal conductivity enhancement in nanotube suspensions. Appl Phys Lett 2001, 79: 2252.
2. Eastman JA, Choi SUS, Li S, Yu W, Thompson LJ: Anomalously increased effective thermal conductivities of ethylene glycol-based nanofluids containing copper nanoparticles. Appl Phys Lett 2001, 78: 718.
3. Lee S, Choi SUS, Li S, Eastman JA: Measuring thermal conductivity of fluids containing oxide nanoparticles. J Heat Transfer 1999, 121: 280.
4. Das SK, Putra N, Thiesen P, Roetzel W: Temperature dependence of thermal conductivity enhancement for nanofluids. J Heat transfer 2003, 125: 567.

5. Wang X, Xu X, Choi SUS: Thermal conductivity of nanoparticle fluid mixture. J Thermophys Heat Transfer 1999, 13: 474.

6. Yang Y, Zhang ZG, Grulke EA, Anderson WB, Wu G: Heat transfer properties of nanoparticle-in-fluid dispersions (nanofluids) in larninar flow. Int J Heat Mass Transfer 2005, 48: 1107.

7. Heris SZ, Etemad G, Esfahany MN: Experimental investigation of oxide nanofluids laminar flow convective heat transfer. Int Commun Heat Mass Transfer 2006, 33: 529.

8. Ding Y, Alias H, Wen D, Williams RA: Heat transfer of aqueous suspensions of carbon nanotubes (CNT nanofluids). Int J Heat Mass Transfer 2006, 49: 240.

9. Mingo N, Broido DA: Carbon Nanotube Ballistic Thermal Conductance and Its Limits. Phys Rev Lett 2005, 95: 096105.

10. Saito K, Nakamura J, Natori A: Ballistic thermal conductance of a graphene sheet. Phys Rev B: Condens Matter 2007, 76: 115409.

11. Balandin AA, Ghosh S, Bao W, Calizo I, Teweldebrhan D, Miao F, Lau CN: Superior Thermal Conductivity of Single-Layer Graphene. Nano Lett 2008, 8: 902.

12. Hummers WS, Offeman RE: Preparation of Graphitic Oxide. J Am Chem Soc 1958, 80: 1339.

13. Kaniyoor A, Baby TT, Ramaprabhu S: Graphene synthesis via hydrogen induced low temperature exfoliation of graphite oxide. J Mater Chem 2010, 20: 8467.

14. Si Y, Samulski ET: Exfoliated Graphene Separated by Platinum Nanoparticles. Chem Mater 2008, 20: 6792.

15. Wei L, Ming TD, Bing HY, Hui YC, Qiang SZ, Cheng CX, Meng CC, Xiang HP, Chang L, Hong YQ: Low-Temperature

Exfoliated Graphenes: Vacuum-Promoted Exfoliation and Electrochemical Energy Storage. ACS Nano 2009, 3: 3730.

16. Li ZQ, Lu CJ, Xia ZP, Zhou Y, Luo Z: X-ray diffraction patterns of graphite and turbostratic carbon. Carbon 2007, 45: 1686.

17. Wu H-L, Yang Y-T, Ma C-CM, Kuan H-C: Molecular Mobility of Free-Radical-Functionalized Carbon-Nanotube/Siloxane/Poly(urea urethane) Nanocomposites. J Polym Sci A: Polym Chem 2005, 43: 6084.

18. Liu M-S, Lin MC-C, Huang I-T, Wang C-C: Enhancement of thermal conductivity with carbon nanotubes for nanofluids. Int Commun Heat Mass Transfer 2005, 32: 1202.

19. Maxwell JC: A Treatise on Electricity and Magnetism. Volume 1. Oxford: Pergamon; 1904.

20. Hamilton RL, Crosser OK: Thermal conductivity of heterogeneous two-component systems. Ind Eng Chem Fundam 1962, 1: 187.

21. Hasselman DPH, Johnson LF: Effective thermal conductivity of composites with interfacial thermal barrier resistance. J Compos Mater 1987, 21: 508.

22. Shah RK: Thermal Entry Length Solutions for the Circular Tube nd Parallel Plates. In Proceedings of the 3rd National Heat Mass Transfer Conference. Volume 1. Indian Institute of Technology, Bombay; 1975. Paper no. HMT-11–75 Paper no. HMT-11-75

23. Dittus FW, Boelter LMK: Heat Transfer in Automobile Radiators of the Tubular Type. University of California Publications in English 1930, 2: 443–461.

24. Xuan YM, Li Q: Investigation on convective heat transfer and flow features of nanofluids. J Heat Transfer 2003, 125: 151.

25. Wen DS, Ding YL: Convective heat transfer of nanofluids at the entrance region under laminar flow conditions. Int J Heat Mass Transfer 2004, 47: 5181.

26. Xuan Y, Li Q, Hu W: Aggregation structure and thermal conductivity of nanofluids. AIChE J 2003, 49: 1038.

8

CHAPTER

EXERGY ANALYSIS OF A GROUND-COUPLED HEAT PUMP HEATING SYSTEM WITH DIFFERENT TERMINALS

Xiao Chen [1,*] and Xiaoli Hao [2]

[1] *Department of Architectural Engineering, Hunan Institute of Engineering, Xiangtan, 411104, China*

[2] *College of Energy and Safety Engineering, Hunan University of Science and Technology, Xiangtan, 411201, China*

ABSTRACT

In order to evaluate and improve the performance of a ground-coupled heat pump (GCHP) heating system with radiant floors as terminals, an exergy analysis based on test results is performed in this study. The system is divided into four subsystems, and the exergy loss and exergy efficiency of each subsystem are calculated using the expressions derived based on exergy balance equations. The average values of the measured parameters are used for the exergy analysis. The analysis results show that the two largest exergy losses occur in the heat pump and terminals, with losses of 55.3% and 22.06%, respectively, and the lowest exergy efficiency occurs in the ground heat exchange system. Therefore, GCHP system designers should pay close attention to the selection of heat pumps and terminals, especially in the design of ground heat exchange systems. Compared with the scenario system in which fan coil units (FCUs) are substituted for the radiant floors, the adoption of radiant floors can result in a decrease of 12% in heating load, an increase of 3.24% in exergy efficiency of terminals and an increase of 1.18% in total exergy efficiency of the system. The results may point out the direction and ways of optimizing GCHP systems.

Keywords: ground-coupled heat pump; exergy analysis; exergy loss; exergy efficiency; radiant floor; fan coil unit

1. INTRODUCTION

Recent years have witnessed the considerable rise of the building energy proportion in the total energy consumption in China. During this period, the utilization of renewable energy in the building sectors of China has increased significantly, especially due to the favorable policies supporting the use of renewable energy. Ground-source heat pumps, including ground-coupled heat pumps, groundwater heat pumps and surface water heat pumps [1], have become increasingly popular in China for heating and cooling of buildings because of their higher energy efficiency compared to conventional air conditioning systems. The total building area of ground-source heat pump (GSHP) applications in China had reached 2.4 × 108 m2 by the end of 2011 [2].

Exergy is defined as the maximum amount of work which can be produced by a system or a flow of matter or energy as it comes to equilibrium with a specified reference environment [3]. The exergy of an energy form or a substance is a measure of its usefulness or quality, and thus is a measure of its potential to cause change [4,5]. Unlike energy, exergy is conserved only during ideal processes and destroyed due to irreversibilities in real processes [6]. Exergy analysis is based on the first and second laws of thermodynamics, and can be used to identify the main sources of irreversibility (exergy loss) and to minimize the generation of entropy in a given process where the transfer of energy and material take place [3,7]. Exergy analysis is widely used for estimation of how near a process is to "thermodynamic ideality", and has become an effective tool for engineers to reveal the potentials to improve energy efficiency by reducing the inefficiencies in energy systems [8,9].

Exergy analysis has been applied to GCHP systems by some researchers [10–14]. Hepbasli and Akdemir [10] presented an energy and exergy analysis of a GCHP system with a fan coil unit (FCU) as terminal. The exergy transfer between the components and the exergy destructions in each component of the GCHP system were determined by the average measured parameters obtained from the experimental results in February 2001. Ozgener and Hepbasli [11] conducted an energy and exergy analysis of a solar-assisted GCHP greenhouse heating system. Exergetic efficiencies and exergy destruction rate of the system components were determined and the potential for improvements was also presented. Esen et al. [12] investigated the energetic and exergetic efficiencies of an experimental GCHP system with two horizontal ground heat exchangers (GHE), which were respectively buried in a 1 m and a 2 m depth trench. A water-to-air heat pump with a capacity of 4.28 kW was used. The results show that the energetic and exergetic efficiencies of the system increase when increasing the buried depth of horizontal GHE. Kuzgunkaya and Hepbasli [13] presented an exergetic assessment of a GCHP drying system in Turkey. The exergy destructions in the overall GCHP system are quantified using the average values of experimentally measured parameters. Exergy efficiencies of the system components are determined to assess their performances and to elucidate potentials for improvement. The total exergy efficiency of the GCHP drying system is estimated to be 15.5% at a reference temperature of 27 °C. Bi et al. [14] presented a comprehensive exergy analysis of a GCHP system with FCU as terminals for both heating and cooling modes. The results show that in the whole system the location of maximum exergy loss ratio is the compressor, while the location of minimum exergy efficiency and thermodynamic perfect degree is the GHE.

As far as other forms of GSHP systems are concerned, Akpinar and Hepbasli [15] investigated the exergetic performance of two types of GSHP systems installed in Turkey based on the actual operational data. The first one is a heat pump system utilizing geothermal water as heat source, while the second one is a GCHP system with vertical GHE. The highest irreversibility of each system occurs in the compressor, and the second largest irreversibility of both systems occurs in the condenser. Balta et al. [16] performed an exergy analysis of a low exergy heating system for a room from the power plant through the heat pump to the building envelope. The heat produced by solar collectors is stored in underground storage tanks, and the heat pump utilizes the water in underground tanks as heat source in heating season. They investigated the energy and exergy flows, and quantified exergy losses in the overall system. Xiang et al. [17] performed an exergetic cost analysis of a space heating system with groundwater heat pumps.

In the literature, exergy analysis has been applied to different types of GSHP systems [10–17]. Previous studies mainly focused on the exergetic performance analysis of system components and GSHP systems, and few on the influence of different terminals on the exergetic performance of system.

In recent years, floor radiant heating systems have come into general application in Northern China due to their thermal comfort and energy savings advantages compared with traditional heating systems. In south China, fan coil units are commonly used as terminals of building heating and cooling systems, and radiant floors are less used as terminals of heating systems. For the sake of rational selection of terminals, this study deals with the influence of the two types of terminals on the energy and exergetic performance of a GCHP heating system. The exergy flows and exergy efficiencies in a GCHP heating system with radiant floors and in a scenario system with FCU as terminal are analyzed respectively. The analysis results are then used to give suggestions for the improvement of the system performance.

2. SYSTEM DESCRIPTION

The GCHP heating system under investigation is located in Changsha, the capital of Hunan Province, which has a typical climate characterized by cold winters and hot summers. According to the meteorological parameter statistics of Changsha in the most recent 30 years, the average number of total days in which daily average temperature was below 5 °C during a year was 48. The average relative humidity in the coldest month was up to 83%, which would result in a more intense feeling of cold. The heat pump system supplies heat to two six-storied residential buildings with a total building area of 5420 m2. The heat transfer coefficients of the walls, windows and roofs are all chosen to satisfy the requirements of the China Industry Standard JGJ 134-2010 [18].

A schematic diagram of the system under investigation is shown in Figure 1. Vertical 1.26 inch nominal diameter U-bend ground heat exchangers are installed. Forty 60 m-depth boreholes are arranged around the two buildings. According to the local climate and experience operating other GCHP applications, the circulating water in the ground loop of GCHP systems in South China does not freeze in winter, therefore, water is used as the circulating fluid in the ground loop. The water in the ground loop is circulated through the evaporator of a water-to-water heat pump equipped with a screw compressor, and returns to the GHE after rejecting heat in the evaporator. The heat pump receives heat from the circulating water in the ground loop, and rejects heat in the condenser to the circulating water in the heat distribution system. The heated circulating water is then pumped to the terminals to heat the rooms.

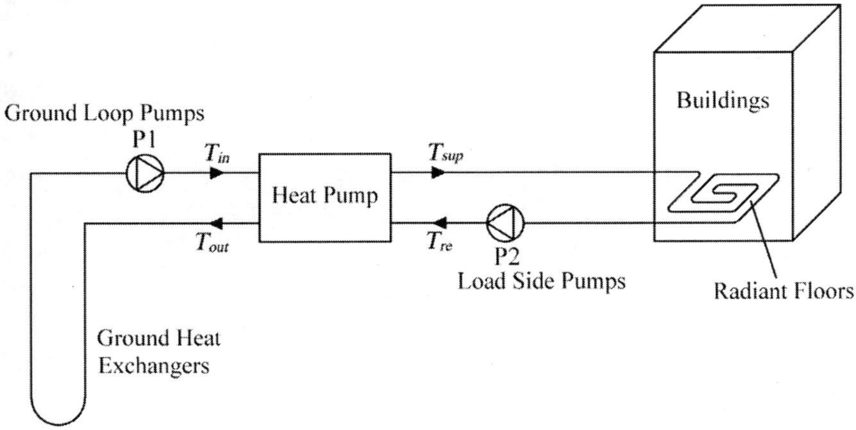

Figure 1. Schematic diagram of the GCHP system.

According to the initial scheme, fan coil units would have been used as terminals for both heating and cooling of the two buildings. However, the buildings in Changsha are generally cooling-dominated and therefore reject more heat than they extract over the annual GCHP system operation cycle. Greater ground heat exchanger lengths would be necessary to adequately dissipate the building heat, which would bring about excessive initial cost over the budget. Therefore, the property developer adopted another scheme in which the GCHP system only operates in winter for heating of rooms, and the dispersed room air conditioners operate in summer for cooling of rooms. Floor radiant heating is used for heating of the two buildings to achieve enhanced thermal comfort.

3. ANALYSIS FOR EXERGY FLOW AND EXERGY EFFICIENCY OF EACH SUBSYSTEM

The GCHP system can be divided into four subsystems, which are respectively the ground heat exchange system, heat pump, heat distribution system and terminals. The exergy flows of each subsystem are illustrated in Figure 2. To carry out the analysis, it is assumed that all processes are steady-state and steady flow with negligible potential and kinetic energy effects and no chemical or nuclear reactions. [10,12,13,15]. For a general steady-state and steady flow process, the exergy balance equations are applied to find the exergy loss of each subsystem [19]. Exergy efficiencies often give more revealing insights into process performance than energy efficiencies, and can provide a measure of potential for improvement [20]. The exergy efficiency of each subsystem is defined as the ratio of exergy product rate to exergy input rate [3].

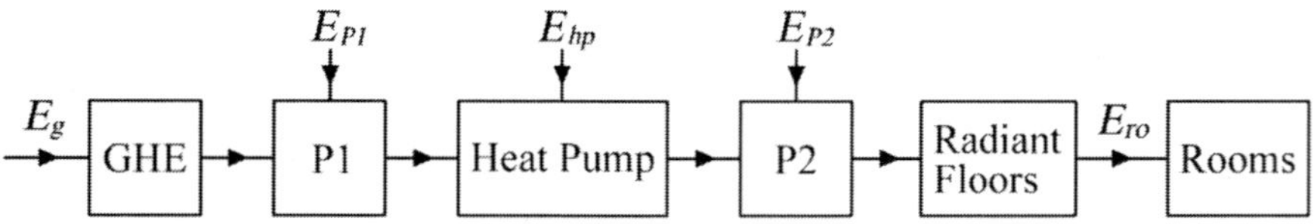

Figure 2. Exergy flows in the GCHP system with radiant floors.

3.1. Ground Heat Exchange System

The ground heat exchange system consists of GHE and ground loop pumps. The exergy input rate from ground can be calculated as:

$$E_g = Q_g\left(1 - \frac{T_0}{T_g}\right) \tag{1}$$

$$Q_g = m_g c_w \left(T_{in} - T_{out}\right) \tag{2}$$

where Eg is the exergy input rate from ground (kW); Qg is the heat extraction rate from ground (kW); T0 is the reference temperature, considered as the outdoor air temperature when the GCHP system operates (K); Tg is ground temperature (K); mg is the mass flow rate of circulating water in ground loop (kg/s); cw is the specific heat of water (kJ/kg·K); Tin and Tout are respectively the inlet and outlet temperatures of the evaporator (K).

The exergy loss of GHE results from the heat exchange between circulating water and ground, and it can be expressed as:

$$I_g = E_g + m_g c_w (T_{out} - T_{in}) - m_g c_w T_0 \ln \frac{T_{out}}{T_{in}} \quad (3)$$

where Ig is the exergy loss of GHE (kW).

The exergy efficiency of ground heat exchange system is calculated as follows:

$$\eta_g = \frac{Q_g - m_g c_w T_0 \ln \frac{T_{in}}{T_{out}}}{Q_g \left(1 - \frac{T_0}{T_g}\right) + E_{P1}} \quad (4)$$

$$E_{P1} = \phi_{P1} W_{P1} \quad (5)$$

where ηg is the exergy efficiency of ground heat exchange system (%); EP1 is the mechanical exergy input rate of ground loop pumps (kW); ϕ P1 is the motor efficiency of ground loop pumps; WP1 is the electric power of ground loop pumps (kW).

3.2. Heat Pump

The exergy loss of a heat pump consists of the exergy losses of all components in a heat pump. The exergy loss of heat pump is calculated as a whole in this analysis, and is given by:

$$I_{hp} = E_{hp} + Q_g - m_g c_w T_0 \ln \frac{T_{in}}{T_{out}} - \left[Q_{hp} - m_h c_w T_0 \ln \frac{T_{sup}}{T_{re}} \right] \quad (6)$$

$$Q_{hp} = m_h c_w \left(T_{sup} - T_{re} \right) \quad (7)$$

where Ihp is the exergy loss of heat pump (kW); Ehp is the mechanical exergy input rate of heat pump (kW); Qhp is the heat output of heat pump (kW); mh is the mass flow rate of hot water (kg/s); Tsup and Tre are respectively the supply and return water temperatures of the condenser (K).

The exergy efficiency of heat pump is calculated by:

$$\eta_{hp} = \frac{Q_{hp} - m_h c_w T_0 \ln \frac{T_{sup}}{T_{re}}}{E_{hp} + Q_g - m_g c_w T_0 \ln \frac{T_{in}}{T_{out}}} \quad (8)$$

$$E_{hp} = \phi_{com} W_{hp}$$

9)

where ηhp is the exergy efficiency of heat pump; ϕ com is the motor efficiency of screw compressor; Whp is the electric power of heat pump (kW).

3.3. Heat Distribution System

The exergy loss of heat distribution results from the heat loss of network, and can be expressed as:

$$I_{hd} = \gamma_{hd} Q_{hp} \left(1 - \frac{T_0}{T_{sup} - T_{re}} \ln \frac{T_{sup}}{T_{re}} \right) \quad (10)$$

where Ihd is the exergy loss of heat distribution (kW); γhd is the heat loss ratio of heat distribution network (%). The total heat distribution distance is as short as 120 m. For simplicity, the heat loss ratio of heat distribution network is taken to be 1.5%.

In view of the small heat loss ratio of heat distribution network, the inlet and outlet water temperatures of terminals are respectively regarded as equal to the supply and return water temperatures of the condenser. The exergy efficiency of heat distribution system is given by:

$$\eta_{hd} = \frac{Q_H \left(1 - \frac{T_0}{T_{sup} - T_{re}} \ln \frac{T_{sup}}{T_{re}} \right)}{Q_{hp} \left(1 - \frac{T_0}{T_{sup} - T_{re}} \ln \frac{T_{sup}}{T_{re}} \right) + E_{P2}} \quad (11)$$

$$Q_H = (1 - \gamma_{hd}) Q_{hp} \quad (12)$$

$$E_{P2} = \phi_{P2} W_{P2} \quad (13)$$

where ηhd is the exergy efficiency of heat distribution system (%); QH is the heating load of buildings (kW); EP2 is the mechanical exergy input rate of load side pumps (kW); ϕ P2 is the motor efficiency of load side pumps; WP2 is the electric power of load side pumps (kW).

3.4. Terminals

The exergy rate transferred to the rooms by terminals is calculated by:

$$E_{ro} = Q_H\left(1-\frac{T_0}{T_a}\right) \tag{14}$$

The exergy loss of terminals is expressed by:

$$I_{ter} = Q_H\left(1-\frac{T_0}{T_{sup}-T_{re}}\ln\frac{T_{sup}}{T_{re}}\right)-Q_H\left(1-\frac{T_0}{T_a}\right) \tag{15}$$

where Iter is the exergy loss of terminals (kW); Ta is indoor air temperature (K).

The exergy efficiency of radiant floors is calculated by:

$$\eta_{rf} = \frac{1-\dfrac{T_0}{T_a}}{1-\dfrac{T_0}{T_{sup}-T_{re}}\ln\dfrac{T_{sup}}{T_{re}}} \tag{16}$$

where ηrf is the exergy efficiency of radiant floors (%).

4. RESULTS AND DISCUSSION

4.1. Test Results

Field tests were performed on 6 January 2013 to acquire operating data for the exergy analysis. The supply and return water temperatures of condenser, the inlet and outlet water temperatures of evaporator, the ground temperature and the air temperature were measured by Pt100 temperature sensors with an uncertainty of ±0.2 °C. The sensor for the measurement of ground temperature is located 30 m under the ground. A paperless recorder based on a single chip microcomputer was used for data acquisition with the aid of temperature sensors.

Water flow rate and electric power were measured, respectively, by ultrasonic flowmeters with an uncertainty of ±1% of actual flow rate and by digital power meters with an uncertainty of ±1% of reading. All parameters were measured when the heat pump operated near steady state condition and near 90% of full load. During the test period, the readings and records fluctuate unobviously. The measured values and calculated values based on measured values for the exergy analysis are listed in Table 1 and Table 2, respectively.

Table 1. The average values of measured parameters.

Measured Parameters	Average Value
Inlet temperature of the evaporator (°C)	9.2
Outlet temperature of the evaporator (°C)	6.1
Supply water temperature of the condenser (°C)	44.2
Return water temperature of the condenser (°C)	40
Flow rate of the circulating water in GHE (m^3/h)	41.6
Flow rate of the hot water (m^3/h)	40.3
Electric power of the heat pump (kW)	58.1
COP (Coeffient of Performance) of the heat pump	3.39
Electric power of the ground loop pumps (kW)	3.8
Electric power of the load side pumps (kW)	4.4
Ground temperature (°C)	16.9
Average outdoor air temperature (°C)	2.8
Average indoor air temperature (°C)	17.4

Table 2. The calculated values based on measured values.

Parameters	Calculated Value
Heat output of heat pump (kW)	195.7
Heat extraction rate from ground (kW)	150
COP (Coeffient of Performance) of the heat pump	3.39
Heating load of the buildings (kW)	192.8
Heat loss rate of heat distribution network (kW)	2.9

4.2. Exergy Losses and Exergy Efficiencies

The equations in Section 3 and test results were used to calculate the exergy loss and exergy efficiency of each subsystem. The motor efficiencies of the screw compressor and water pumps are respectively 88% and 85%. The exergy rates and their percentages are presented using the Grassmann diagram in Figure 3. The exergy input rate from ground is 7.29 kW. Total exergy input rate consists of the exergy input rate from ground, the mechanical exergy input rates of the heat pump and all water pumps. The largest exergy loss and exergy input rate all occur in the heat pump. The second largest exergy loss occurs in floor heating process. More attention should be paid to the selection of heat pumps and terminals. In view of heat pumps operate under part-load conditions at most time, the heat pumps with relative high COP value under part-load conditions should be preferred.

The exergy efficiencies of the four subsystems are listed in Table 3. The exergy efficiency of the ground heat exchange system is the lowest among the four subsystems. Improving the effectiveness of GHE and reducing the pumping energy for fluid circulation in GHE are all effective measures to improve the exergy efficiency of ground heat exchange systems. In some GCHP heat pump systems, pumps with higher pumping head than needed operate at high flow rate, low temperature difference and low pump efficiency. A proper pumping head should be selected during the system design stage. In addition, variable flow operation of pumps is necessary under part-load operating conditions.

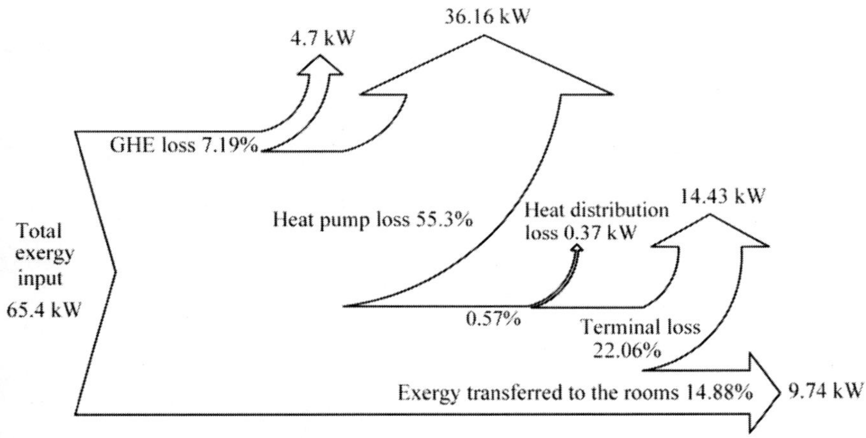

Figure 3. Grassmann diagram of the GCHP system.

Table 3. Exergy efficiencies of the four subsystems.

Subsystems	Exergy Efficiency (%)
Ground heat exchange system	24.63
Heat pump	45.66
Heat distribution system	85.47
Terminals (Radiant floors)	40.31

The total exergy efficiency of the GCHP heating system is 14.88% at a reference temperature of 2.8 °C. The total exergy efficiency of the system and COP of the heat pump are compared with those of a GCHP system with horizontal GHE buried in a 1 m depth trench in Ref. [12] and a GCHP system with vertical GHE in Ref. [14], as shown in Table 4. In view of the fact the reference temperatures in Ref. [12] and Ref. [14] are 1 °C and 0 °C, respectively, the total exergy efficiency of the system in this study is also calculated based on the reference temperatures of 1 °C and 0 °C for better comparison.

Table 4. Comparisons of total exergy efficiency and COP of heat pump.

	COP of Heat Pump	Total Exergy Efficiencies at Different Reference Temperature (%)		
		2.8 °C	1 °C	0 °C
Ref. [12]	2.5		53.1	
Ref. [13]	3.9			10.0
This study	3.39	14.88	16.5	17.3

As seen from Table 4, the total exergy efficiency decreases with the increase of reference temperature. Although the COP of the heat pump in Ref. [12] is lower than that of any other system in Table 4, the total exergy efficiency of the system in Ref. [12] is 36.6% higher than that of the system in this study. Unlike the water-to-water heat pumps in this study and Ref. [14], the heat pump in Ref. [12] is a small water-to-air heat pump. The condenser fan blows across the condenser to disperse the warmed air into the room, which eliminate the existence of terminals and load side pumps. Thus, the total exergy input and total exergy losses are decreased obviously. The total exergy efficiency in this study is 7.3% higher than that of the system in Ref. [14], which can be attributed to relatively higher pumping efficiency and the adoption of radiant floors in this study. The ratio of pumping energy to the heat transferred by pumps (REH) is usually used to evaluate pumping efficiency [21]. The higher REH is, the lower pumping efficiency. The ratio of load side pumping energy to the heat transferred by load side pumps in this study is

0.022, and that in Ref. [14] is up to 0.041. The terminals of the system in Ref. [14] are FCU. The exergy efficiency of FCU is lower than that of radiant floors, which will be discussed in Section 4.3.2 detailly.

4.3. Comparisons between Radiant Floors and FCU

4.3.1. Comparison of the Energy Performance

The energy-saving benefit resulting from the adoption of radiant floors lies in the decrease of heating load. For the rooms with radiant floors, the same body feeling temperature can be achieved even though indoor air temperature is lower than the rooms with other types of terminals. According to China National Standard, GB 50736-2012 [21], the design temperature of the rooms with radiant floors should be 2 °C lower than that of the rooms with other types of terminals. Thus, the indoor air temperature of the rooms with FCU is taken as 19.4 °C. The corresponding decrease of heating load can be deduced from the calculation method of heating load as follows:

$$Q_H = \sum_i K_i U_i A_i \left(T_a - T_0\right) + c_a \rho \left(T_a - T_0\right) \frac{NV}{3600} \quad (17)$$

where Ki is temperature correction factors; Ui is the heat transfer coefficient of building envelope (kW/m2·K); Ai is the area of building envelope (m2); N is air exchange rate of rooms (times per hour); ca is the specific heat of air (kJ/kg·K); ρ is air density (kg/m3); V is interior volume of rooms (m3). Above equation indicates that heating load is proportional to indoor and outdoor temperature difference. For the same two buildings in this case, the decrement rate of heating load (DRHL) resulting from the adoption of radiant floors can be calculated as follows:

$$DRHL = \frac{(19.4 - 2.8) - (17.4 - 2.8)}{19.4 - 2.8} = 12\% \quad (18)$$

4.3.2. Comparison of the Exergetic Performance

To evaluate the exergy efficiency of FCU, it is necessary to find the relationship between the electric power and heat transfer rate of FCU. According to the manufacturer's catalog data [22], the relationship between the electric power and heat transfer rate of FCU is plotted in Figure 4.

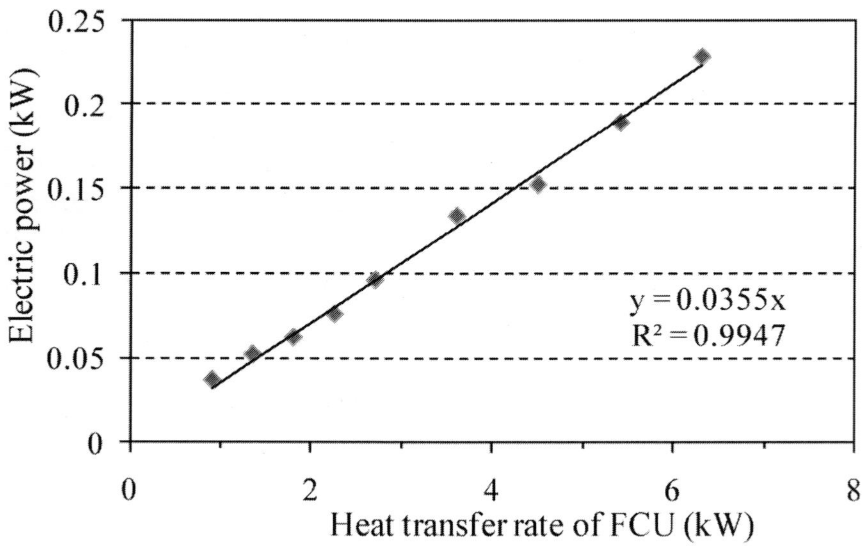

Figure 4. Electric power versus heat transfer rate of FCU.

The electric power of FCU is generally linearly related to the heat transfer rate of FCU, and can be calculated by following fitting formula:

$$W_{FCU} = 0.0355Q_{FCU} \qquad (R^2 = 0.9947) \tag{19}$$

where WFCU is the electric power of FCU (kW); QFCU is the heat transfer rate of FCU (kW).

If fan coil units were adopted as terminals in this GCHP system, the exergy efficiency of FCU could be calculated by:

$$\eta_{FCU} = \frac{1-\dfrac{T_0}{T_a}}{1-\dfrac{T_0}{T_{sup}-T_{re}}\ln\dfrac{T_{sup}}{T_{re}}+0.0355\phi_{FCU}} \quad (20)$$

where ηFCU is the exergy efficiency of FCU (%); ϕ FCU is the motor efficiency of FCU and equal to 80%. Suppose that the values of T0, Tsup and Tre are equal to the test results, the exergy efficiency of FCU will be 37.07%. The exergy efficiency of radiant floors is 3.24% higher than that of FCU due to the fact that the radiant floors operate without electric power. If FCU were substituted for the radiant floors in this GCHP heating system in terms of the test results, the total exergy efficiency of system would drop to 13.7%. Besides the advantages in thermal comfort and energy saving compared with FCU, the adoption of radiant floors can bring a certain increase in the total exergy efficiency of system.

5. CONCLUSIONS

An exergy analysis of a GCHP system with radiant floors as terminals is performed based on test results. The following concluding remarks can be extracted from this study:

The heating load and exergy efficiency of the GCHP heating system with radiant floors are compared with those of the scenario system in which FCU are substituted for the radiant floors. The comparison results show that the adoption of radiant floors can result in a decrease of 12% in heating load, an increase of 3.24% in exergy efficiency of terminals and an increase of 1.18% in total exergy efficiency of system.

The analysis results indicate that the largest exergy loss occurs in heat pump, and the second largest exergy loss occurs in terminals. The ground heat exchange system has the lowest exergy efficiency among the four subsystems. Designers of GCHP systems should pay close attention to the selection of heat pumps and terminals, especially to the design of ground heat exchange systems.

Both load side pumping energy and source side pumping energy have obvious influence on the energy and exergetic performance of a GCHP heating system. The total exergy efficiency of a GCHP heating system will decrease with the increase of reference temperature.

ACKNOWLEDGMENTS

This work was financially supported by Hunan Provincial Natural Science Foundation of China (12JJ9013) and National Natural Science Foundation of China (51276058).

NOMENCLATURE

Ai area of building envelope (m2)

c specific heat (kJ/kg·K)

E exergy (kW)

I exergy loss (kW)

Ki temperature correction factors

m mass flow rate (kg/s)

N air exchange rate of rooms (times per hour)

Q heat transfer rate (kW)

T temperature (K)

Ui heat transfer coefficient of building envelope (kW/m2·K)

V interior volume of rooms (m3)

W electric power (kW)

Greek Letters

η exergy efficiency (%)

ϕ motor efficiency (%)

ρ air density (kg/m3)

γ heat loss ratio (%)

Subscripts

a air

com screw compressor

FCU fan coil units

g ground

H	heating
h	hot water
hd	heat distribution
hp	heat pump
in	inlet of the evaporator
out	outlet of the evaporator
P1	ground loop pumps
P2	load side pumps
re	return water
rf	radiant floor
ro	rooms
sup	supply water
ter	terminals
w	water
0	reference environment condition

AUTHOR CONTRIBUTIONS

Xiao Chen and Xiaoli Hao contributed to the conception of the study, the development of the methodology for exergy analysis and the field tests. Xiao Chen performed the data analyses and wrote the manuscript. Both authors have read and approved the final manuscript.

REFERENCES

1. American Society of Heating, Refrigerating and Air-conditioning Engineers (ASHRAE). 2011 ASHRAE Handbook-HVAC Applications; ASHRAE: Atlanta, GA, USA, 2011.
2. Qiu, B.X. Situation and task of developing green and energy-efficient building in China. Proceedings of the 8th International Conference on Green and Energy-efficient Building, Beijing, China, 28–31 March 2012.
3. Kotas, T.J. The Exergy Method of Thermal Plant Analysis; Paragon Publishing: Trowbridge, UK, 1995.
4. Rosen, M.A.; Dincer, I. Exergy analysis of waste emissions. Int. J. Energy Res. 1999, 23, 1153–1163.
5. Rosen, M.A.; Dincer, I. Thermoeconomic analysis of power plants: An application to a coal-fired electrical generating station. Energy Convers. Manag. 2003, 44, 2743–2761. [Google Scholar]
6. Rosen, M.A.; Dincer, I.; Kanoglu, M. Role of exergy in increasing efficiency and sustainability and reducing environmental impact. Energy Policy 2008, 36, 128–137.

7. Rosen, M.A. Using exergy to correlate energy research investments and efficiencies: Concept and case studies. Entropy 2013, 15, 262–286.
8. Dincer, I.; Cengel, Y.A. Energy, entropy and exergy concepts and their roles in thermal engineering. Entropy 2001, 3, 116–149.
9. Hammond, G.P. Industrial energy analysis, thermodynamics and sustainability. Appl. Energy 2007, 84, 675–700.
10. Hepbasli, A.; Akdemir, O. Energy and exergy analysis of a ground source (geothermal) heat pump system. Energy Convers. Manag. 2004, 45, 737–753.
11. Ozgener, O.; Hepbasli, A. A parametrical study on the energetic and exergetic assessment of a solar-assisted vertical ground-source heat pump system used for heating a greenhouse. Build. Environ. 2007, 42, 11–24.
12. Esen, H.; Inalli, M.; Esen, M.; Pihtili, K. Energy and exergy analysis of a ground-coupled heat pump system with two horizontal ground heat exchangers. Build. Environ. 2007, 42, 3606–3615.
13. Kuzgunkaya, E.H.; Hepbasli, A. Exergetic performance assessment of a ground-source heat pump drying system. Int. J. Energy Res. 2007, 31, 760–777.
14. Bi, Y.H.; Wang, X.H.; Liu, Y. Comprehensive exergy analysis of a ground-source heat pump system for both building heating and cooling modes. Appl. Energy 2009, 86, 2560–2565.
15. Akpinar, E.K.; Hepbasli, A. A comparative study on exergetic assessment of two ground-source (geothermal) heat pump systems for residential applications. Build. Environ. 2007, 42, 2004–2013.

16. Balta, M.T.; Kalinci, Y.; Hepbasli, A. Evaluating a low exergy heating system from the power plant through the heat pump to the building envelope. Energy Build. 2008, 40, 1799–1804.

17. Xiang, J.Y.; Zhao, J.; Qin, N. Exergetic cost analysis of a space heating system. Energy Build. 2010, 42, 1987–1994.

18. China Industry Standard. Design Standard for Energy Efficiency of Residential Buildings in Hot Summer and Cold Winter Zone, JGJ 134–2010; China Architecture and Building Press: Beijing, China, 2010.

19. Cengel, Y.; Boles, M.A. Thermodynamics: An Engineering Approach, 4th ed; McGraw-Hill: New York, NY, USA, 2001.

20. Hernández-Román, M.Á.; Manzano-Ramírez, A.; Pineda-Piñón, J.; Ortega-Moody, J. Exergetic and thermoeconomic analyses of solar air heating processes using a parabolic trough collector. Entropy 2014, 16, 4612–4625.

21. China National Standard. Design Standard for Heating Ventilation and Air Conditioning of Civil Buildings, GB 50736–2012; China Architecture and Building Press: Beijing, China, 2012.

22. Carrier China Limited. Manufacturer Data of Fan Coil Units; Shanghai, China, 2011.

9 CHAPTER

EXPERIMENTAL INVESTIGATION OF COPPER-GROOVED MICRO HEAT PIPES (MHPS)

Xiao-Dong Wang[1], Liang-Liang Zou[2], Ji-Guang Liu[2], Yi Luo[1]*, Gang Liu[2] and Bei-Ke Yu[2]

(1)Key Laboratory for Precision and Non-traditional Machining Technology of Ministry of Education, Dalian University of Technology

(2)Key Laboratory for Micro/Nano Technology and System of Liaoning Province, Dalian University of Technology

ABSTRACT

With the excellent thermal conductivity and the compatibility to micro electromechanical systems technology, silicon is widely used in micro heat pipes (MHPs). Copper shows higher heat transfer capability and capillary traction than silicon. Copper microgrooves were fabricated on the silicon wafer using electroforming technique in this paper. Water contact angle measurements and thermal behavior tests demonstrated that copper-grooved MHPs showed better performance than silicon ones. Under the input power of

5.99 W, the equivalent thermal conductivities of copper-grooved and silicon-grooved MHPs were 228.98 W/K·m and 196.26 W/K·m. This work showed the feasibility of copper grooved silicon based MHPs in heat transfer for high-power light emitting diode (HP LED).

Keywords: MHPs Copper microgrooves Contact angle Thermal behavior

BACKGROUND

In recent years, how to reduce the high-power light emitting diode's (HP LED) thermal resistance, junction temperature and hot spot influence have been studied by different research groups in passive and active cooling mechanism. There are several aspects having been studied in heat dissipation, such as package design [1]–[4], thermal interface material [5], low thermal resistance heat sink material [6]–[9], and cooling systems [10]–[14]. The best-known devices for effective heat transfer or heat spreading with the lowest thermal resistance are heat pipes with vapor chambers which are two-phase heat transfer devices with excellent heat spreading and heat transfer characteristics [15],[16].

Silicon is selected as the ideal material of micro heat pipes (MHPs). To enhance the thermal behavior of MHPs, researchers have introduced a series of methods. Ye H [17] fabricated a fluid channel by wet etching and wafer bonding with silicon and glass. The phase transition of coolant allowed the package temperature to remain below 115°C with LED power up to 2.8 W. Dean R. N [18] presented a MHP constructed from a micromachined base and lid that were bonded together after charging. The base and the lid were fabricated on separate silicon wafers utilizing convension MEMS fabrication technique. The base silicon wafer had 22 channels, where each channel was 100 μm wide and 9.5 mm long. The water-filled MHP had a thermal conductivity of approximately 290 W/ m · K while the Hg-filled MHP had a thermal conductivity of approximately 790 W/ m · K. However, since Hg forms health and safety concerns, this working fluid should be used carefully. A MHP consisted of two layers was introduced by Liu X. W [19]. The bottom layer was Si. Its V-shaped grooves were developed by bulk Si technology. Glass was the up layer, which served as a chamber connecting the vapor phase of all the grooves. The size of the long MHP was 14 mm × 34 mm and the short one was 14 mm × 24 mm. And for 5 W input power, when the filling ratio was 35%, long MHP had the lowest evaporator surface temperature of 90°C. And for the short one, the lowest evaporator surface temperature was 86°C at the optimal filling ratio. Liu W. T [20] fabricated the MHP array in the plate composed of 17 parallel triangular pipes. Nine of the pipes were 371 μm wide and 20 mm long, and the other eight smaller pipes, which served as arteries between every neighbor MHPs, were 268 μm wide and 20 mm long. Ethanol was selected as the working fluid. The working temperature range of MHP was 37.3°C –44.1°C.

Though silicon was widely used in MHPs, it showed poorer surface wettability which was identified as one of the key factors affecting phase change heat transfer [21], compared to the material of copper. Mahmood R. S. Shirazy et al. [22] studied the capillary and wetting characteristics of copper metal foam which was used as a wick in flat heat pipes. Lim H. T, et al. [23] fabricated a copper MHP that worked under adverse-gravity conditions. The wick structure of the flat MHP consisted of fan-shaped microgrooves with a width and depth of about 100 μm and 200 μm, respectively. The fabrication of microgrooves was done using a laser micromachining technique and water was used as the working fluid. Fan-shaped microgrooves were found to induce a greater capillary pressure than triangular microgrooves of a similar size. Subsequent test results confirmed that despite its small size, $56 \times 8 \times 1.5$ mm^3, the FMHP had a high heat transport capacity. The test results demonstrated that a FMHP with fan-shaped microgrooves had a good cooling capacity and worked with no loss of performance under adverse-gravity orientation. Paiva K. V. [24] presented a copper mini flat heat pipe fabricated using the welding diffusion technique. When a plate and a cylinder touched each other, a very sharp edge between them, which could work as a groove, were observed. The mini heat pipe studied in the present paper had $100 \times 30 \times 2\ mm^3$ of dimensions and was made of ten parallel cylindrical copper wires welded between two thin copper sheets each with 0.3 mm of thickness. Water was chosen as the working fluid, and the maximum heat transfer transported by grooves was 1.25 W.

Meng K. [25] identified that Metallic surfaces with high surface free energy were intrinsically hydrophilic, and the performance of MHP could benefit from enhanced wettability. Copper had excellent thermal conductivity and was usually used as heat exchanger materials, but the copper surface was not hydrophilic, which impacted the heat transfer.

The static contact angle of water droplets on air-exposed copper often exceeds 70°. Since wettability evaluated by water contact angle (WCA) is governed mainly by both surface geometrical structures and chemical compositions [26], researchers had put forward efforts to improve the hydrophobicity of copper. Zhang Q. Y. [27] utilized silica hydrophilic and moisture retention property. Considering that the silica cannot be separated from the film, organic resin was added as a binder, and both interaction could play a hydrophilic function. Min J. [28] immersed the copper-finned evaporator in a NaOH and $K_2S_2O_8$ mixed solution for a certain length of time. The hydrophilically-treated evaporator tended to yield a greater cooling capacity than the untreated evaporator. Min J.C. [29] investigated the wetting and corrosion characteristics of hot-water-treated aluminum and copper fin stocks and found that the hot water soak could improve the surface wettability of both aluminum and copper surfaces. Nam Y. [30] reported the heat transfer performance of super hydrophilic Cu micro post wicks fabricated on thin silicon substrates using electrochemical deposition and controlled chemical oxidation. Copper oxide nano-structures formed on the micro post surfaces significantly enhanced the critical heat flux without compromising the effective heat transfer coefficient.

Literatures prove that the copper-grooved silicon wafer can show better heat transfer characteristic than the silicon grooved one. In this paper, a copper-grooved MHP was fabricated using electroforming technique. Water contact angle and the thermal behavior of both copper-grooved and silicon-grooved MHPs were tested.

METHOD

The dimension of the MHPs was designed to suit for 10 W LED module heat dissipation. The dimension of MHP depended on the dimension of the module. The length and width of LED module in this paper were both 16 mm. The overall dimension of MHP was 45 mm (length) × 16 mm (width) × 1.5 mm (thickness). The working area was 35 mm × 10 mm, including condenser, adiabatic and evaporation section. The microgrooves were fabricated on the silicon wafer, which was bonded to Pyrex 7740 glass with a steam chamber. According to the previous study, the steam chamber enhanced the thermal behavior of MHPs compared with the MHPs without the steam chamber [31]. The Pyrex 7740 glass was used to achieve the visualization of how the working fluid was moving.

Silicon heat pipes contained four sections of parallel microgrooves, and the specific dimensions of this type of MHP were shown in Table 1 and Figure 1.

Table 1. The mhps' design dimension

Section	*A*	*B*	*C*	*D*	*E*	*F*	G
Dimension	200 µm /60 µm	180 µm /80 µm	160 µm /100 µm	140 µm /120 µm			

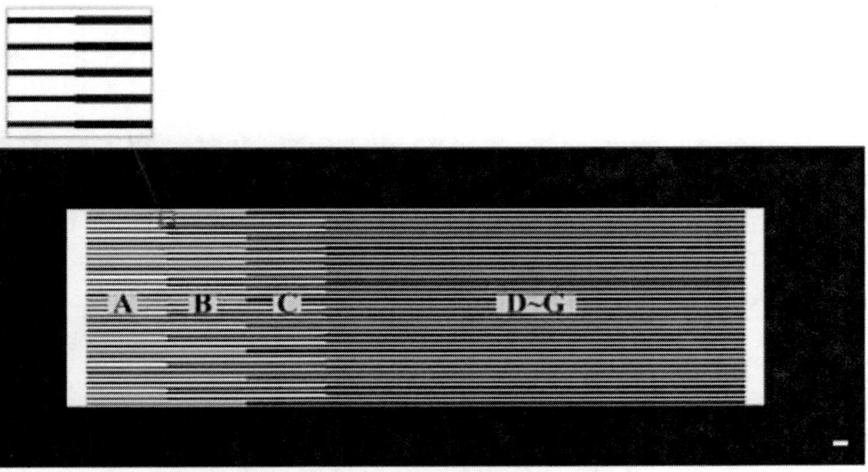

Figure 1. Photolithographic mask of mhp grooves.

In the table above, section A ~ G divided the working area of MHP in the axial. The value before '/' was the width of the microgrooves, and the value after '/' was the space between the microgrooves.

The structure and dimension of the copper-grooved MHP were the same with those of the silicon MHP.

Anisotropic wet etching was used to fabricate silicon microgrooves on the n-type silicon (100) wafer. The depth of the microgrooves was about 200 µm. The Pyrex 7740 was wet etched to form a vapor chamber, and the depth of the vapor chamber was 200 µm. The silicon wafer and the Pyrex 7740 were bonded using anodic bonding technique at the temperature of 450°C and the voltage of 1200 V.

Electroforming process was used to fabricate copper microgrooves on the n-type silicon (100) wafer. The depth of the microgrooves was about 100 μm. In the electroforming process, the current density was 2A/dm^2, and electroforming solution was mixed by the mix of $CuSO_4$ (70 g/l) and H_2SO_4 (12%). The whole process was under the condition of 24°C. Finally, the fabricated copper micro-grooved silicon wafer and Pyrex 7740 glass were enveloped using sealant (Loctite 326 structural adhesive).

Double air pumping charge method was used for the MHP's charging and sealing after the microgrooves were fabricated. This method has been detailed in another paper [32]. The working fluid was degassed DI water, and the filling ratio was about 30%. The fabricated MHPs are shown in Figure 2.

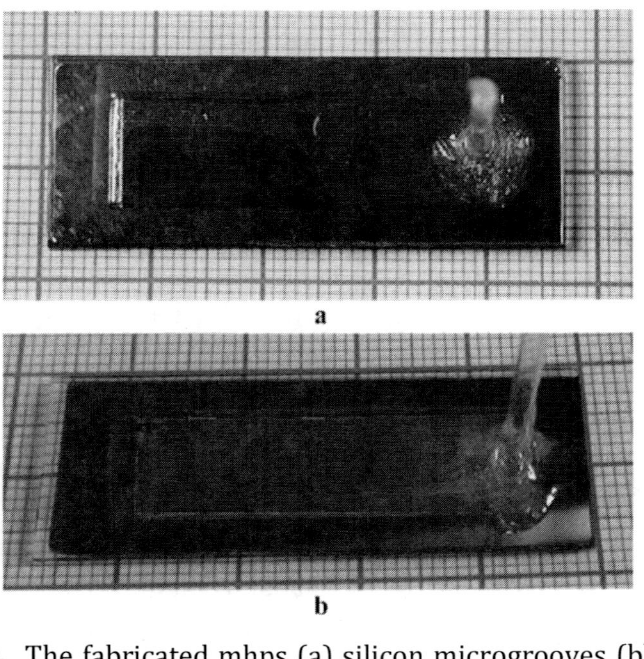

Figure 2. The fabricated mhps (a) silicon microgrooves (b) copper microgrooves.

An experiment was conducted to measure the hydrophobicity between the fabricated silicon microgrooves and copper microgrooves. The drop shape analyzer measuring instrument (DSA 100 Kruss, Germany) was used to measure the contact angle of the fabricated microgrooves. Each MHP was measured by three points (point 1, 2 and 3) which were the central points of the condensation, adiabatic and evaporation section, respectively, as shown in Figure 3. The measured contact angles are shown in Table 2 and Figure 4.

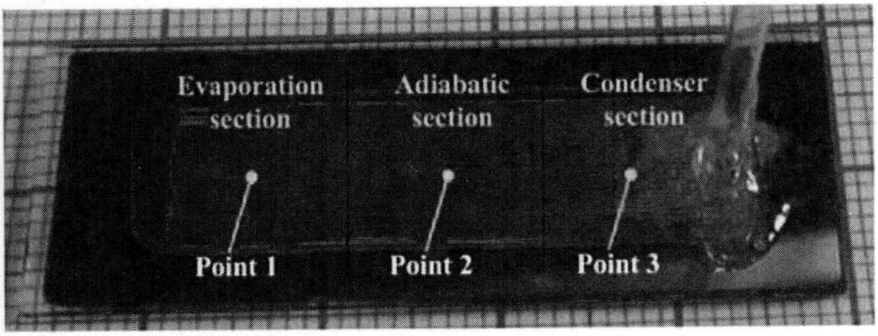

Figure 3. The schematic of three measured points (1, 2 and 3) in copper-grooved silicon base.

Table 2 .the contact angle of grooved silicon and copper

	1	*2*	*3*
Silicon microgrooves	88.2°	88.5°	103.5°
Copper microgrooves	54.8°	53.0°	65.5°

The value showed that the 3 points contact angles of copper microgrooves were smaller than those of silicon microgrooves, respectively. The main reason was the dimension of the micro grooves at these three sections are different, and we wish the condenser section has higher capillary force, that is more hydrophilic, which can help the working fluid traction back to the evaporation section. Since the capillary traction was proportional to the cosine value of the contact angle, copper microgrooves had stronger hydrophilicity and better capillary traction compared with the silicon microgrooves.

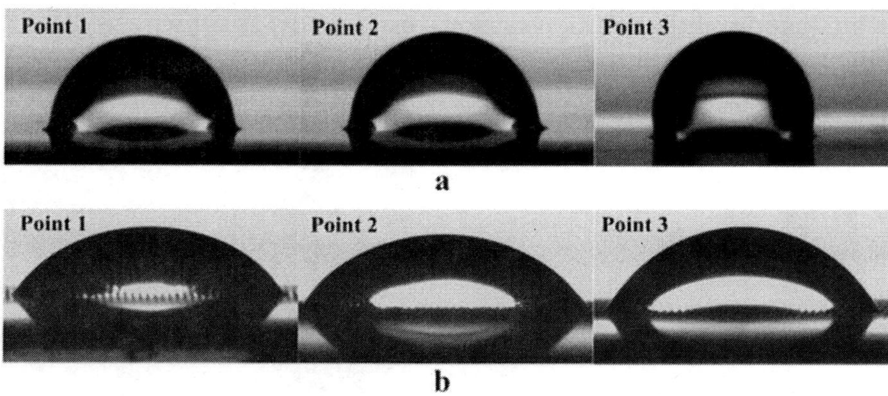

Figure 4. The contact angle of the microgrooves (a) silicon grooves, (b) copper grooves.

The thermal performance of the MHPs can be evaluated by the equivalent thermal conductivity which is calculated as follows:

$$K = \frac{Q_{in}}{T_1 - T_2} \times \frac{L}{A_c} \tag{1}$$

where Q_{in} was the heat input, T_1 and T_2 were the temperatures of the evaporation part and the condensation part, respectively, L was the length between evaporation part and condensation part, and A_c was the cross sectional area of the MHPs.

The thermal behavior testing system was described in another paper [33]. In this system, Two K type thermocouples (Chal-0005, Omega Company, USA) were used to test the temperature at the evaporation and condensation section, respectively. A heat flux sensors (HFS-3,

Omega Company, USA) was used to record the input power of the MHPs.

The accuracy grade of the thermocouple instrument (Testo 735, Germany) was ±0.3°C. Calibration of the thermocouples was done by measuring the temperature change of thermocouples upon the temperature change inside a well-controlled oven. This testing system was used to test the fabricated MHPs, the silicon wafer and the copper wafer. The 10 W LED module was placed as the heat source of the MHP. The voltage and current applied to the LED module was 31.50 V and 0.19 A, respectively, which means the input power was 5.99 W. The temperature rising curves of the two thermocouples at the evaporation and condensation section are shown in Figure 5. A silicon wafer and a copper wafer with the same dimension of the MHP were also tested, respectively.

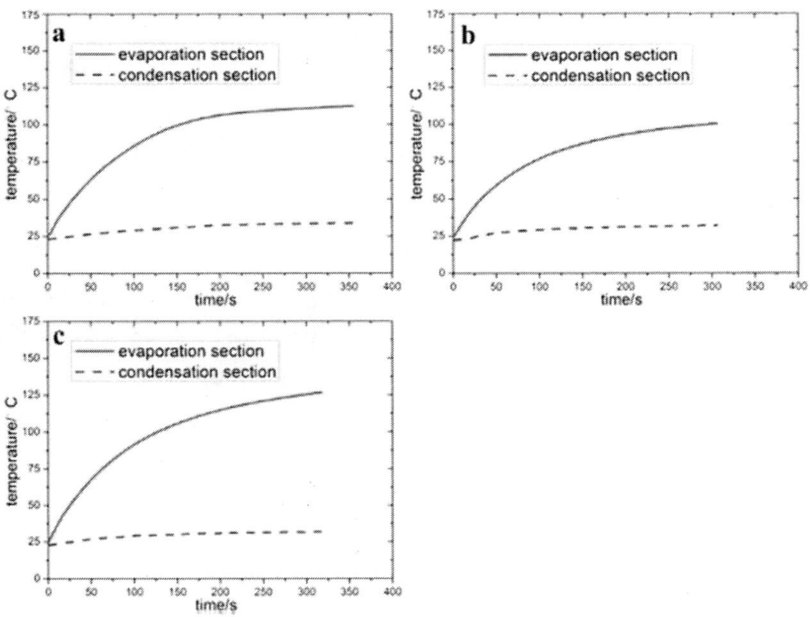

Figure 5. The temperature curve of the mhps. **(a)** and **(b)** were the temperature curves of the silicon MHP and copper MHP, respectively, and **(C)** was the temperature curves of the silicon wafer.

RESULTS AND DISCUSSION

With Equation (1) and the data shown in Figure 5, the equivalent thermal conductivities of the MHPs were calculated and listed in Table 3. As a comparison, the equivalent thermal conductivity of the silicon wafer was also tested, calculated and listed in Table 3.

Table 3. The equivalent thermal conductivity

MHP	K^a (W/K·m)
Silicon microgrooves	196.26
Copper microgrooves	228.98
Si	72.57

[a]*K* is the equivalent thermal conductivity of MHPs and Si wafer.

It is obvious that the equivalent thermal conductivity of MHPs is higher than that of the silicon wafer. The MHPs worked effectively. The equivalent thermal conductivity of the copper-grooved MHP was 2.16 times higher than that of silicon wafer which is the raw material of the MHPs.

The equivalent thermal conductivity of the copper-grooved MHP was higher than that of the silicon MHP. The performance was promoted about 17% only because the grooved material changed from silicon to copper, while the base material was same. The possible reason is that the copper grooves were more hydrophilic, thus higher capillary force could be generated and working fluid could be drawn back to evaporation section more efficiently.

CONCLUSIONS

A copper-grooved silicon MHP was introduced in this paper, and the rectangular-ambulatory structures was designed and fabricated between copper and silicon. Since the copper microgrooves had a smaller contact angle, as well as better capability of capillary traction, the copper-grooved MHPs showed a 17% higher equivalent thermal conductivity than that of silicon-grooved MHPs.

The copper-grooved silicon based MHP was proved to be feasible in heat transfer for HP LEDs. Moreover, future work will be focused on structure optimization to achieve a more efficient and reliable heat transfer.

ACKNOWLEDGMENTS

The authors wish to gratefully acknowledge the financial support provided for this study by the State Key Development Program for Basic Research of China (Grant No. 2011CB013105).

AUTHORS' CONTRIBUTIONS

XDW conceived of the study and proposed the copper-grooved MHP. LLZ fabricated the silicon based MHP and drafted the manuscript. JGL participated in the design of the study and fabricated the copper-grooved MHP. YL conceived of the thermal behavior test system and the structure of the MHPs. GL. performed the thermal behavior test and statistical analysis. BKY participated in MHP design and helped to draft the manuscript. All authors read and approved the final manuscript.

REFERENCES

1. Cheng T, Luo XB, Huang S, Liu S: Thermal analysis and optimization of multiple LED packaging based on a general solution. Int J Therm Sci 2010, 49: 196–201.

2. Juntunen E, Tapaninen O, Sitomaniemi A, Heikkinen VE: Effect of phosphor encapsulant on the thermal resistance of a high-power COB LED module. Trans Compon Packag Manuf Tech 2013, 3: 148–1154.
3. Luo XB, Fu X, Chen F, Zheng H: Phosphor self-heating in phosphor converted light emitting diode packaging. Int J Heat Mass Transf 2013, 58: 276–281.
4. Hu R, Luo XB, Zheng H: Hotspot location shift in the high power phosphor converted white light-emitting diode package. Jpn J Appl Phys 2012, 51: 1–4. 09MK05 09MK05
5. Liou BH, Chen CH, Horng RH, Chiang YC, Wuu DS: Improvement of thermal management of high-power GaN-based light-emitting diodes. Microelectron Reliab 2012, 52: 861–865.
6. Sim JK, Ashok K, Ra YH, Im HC, Baek BJ, Lee CR: Characteristic enhancement of white LED lamp using low temperature co-fired ceramic-chip on board package. Curr Appl Phys 2012, 12: 494–498.
7. Kang M, Kang S: Influence of Al2O3 additions on the crystallization mechanism and properties of diopside/anorthite hybrid glass-ceramics for LED packaging materials. J Cryst Growth 2011, 326: 124–127.
8. Zhou W, Qi S, Li H, Shao S: Study on insulating thermal conductive BN/HDPE composites. Thermochim Acta 2007, 452: 36–42.
9. Luo X, Mao Z: Thermal modeling and design for microchannel cold plate with high temperature uniformity subjected to multiple heat sources. Int Commun Heat Mass Transf 2012, 39: 781–785.

10. Kim L, Choi JH, Jang SH, Shin MW: Thermal analysis of LED array system with heat pipe. Thermochim Acta 2007, 455: 21–25.

11. Lu XY, Hua TC, Wang YP: Thermal analysis of high power LED package with heat pipe heat sink. Microelectron J 2011, 42: 1257–1262.

12. Wan ZM, Liu J, Su KL, Hu XH MSS: Flow and heat transfer in porous micro heat sink for thermal management of high power LEDs. Microelectron J 2011, 42: 632–637.

13. Liu S, Yang JH, Gan ZY, Luo XB: Structural optimization of a microjet based cooling system for high power LEDs. Int J Therm Sci 2008, 47: 1086–1095.

14. Anithambigai P, Dinashi K, Mutharasu D, Shanmugan S, Lim CK: Thermal analysis of power LED employing dual interface method and water flow as a cooling system. Thermochim Acta 2011, 523: 237–244.

15. Kosakabe T, Mochizuki M, Mashiko K, Saito Y, Kiyooka F, Horiuchi Y, Cabusao G, Nguyen T: Heat spreader technology for silicon chip. 2010 IEEE CPMT Symposium Japan ICSJ10 2010, doi:10.1109/CPMTSYMPJ.2010.5680370.

16. Hirachan A, Agonafer D: Removing the Hot-Spots in High Power Devices Using the Thermoelectric Cooler and Micro Heat Pipe. 2012.

17. Ye H, Mihailovic M, Wong CKY, Zeijl HW, Gielen AWJ, Zhang GQ: Two-phase cooling of light emitting diode for higher light output and increased efficiency. Appl Therm Eng 2013, 55: 353–359.

18. Robert ND, Daniel KH, Ashish YP, Gary DW: Liquid metal-filled micro heat pipes for thermal management of solid-state devices. IEEE Trans Ind Electron 2012, 59(12):4888–4894.

19. Liu XW, Wang C, Wang LW, Han T: Design and fabrication of flat heat pipes with different length. 2011 Acad Int Sym Optoelectron Microelectron Tech (AISOMT) 2011. doi:10.1109/AISMOT.2011.6159373.

20. Liu W, Kang J, Fu X, Stefanini C, Dario P: Analysis on heat resistance of the micro heat pipe with arteries. Microelectron Eng 2011, 88: 2255–2258.

21. Nam Y, Ju YS: A comparative study of the morphologyand wetting characteristics of micro/nanostructured Cu surfaces for phasechange heat transfer applications. J Adhes Sci Technol 2012, 27: 2163–2176.

22. Shirazy MRS, Frechette LG: Capillary and wetting properties of copper metal foam in the presence of evaporation and sintered walls. Int J Heat Mass Transf 2013, 58: 282–291.

23. Lim HT, Kim SH, Im HD, Oh KH, Jeong SH: Fabrication and evaluation of a copperflat micro heat pipe working underadverse-gravity orientation. J Micromech Microeng 2008, 18: 105013(1–8).

24. Paiva KV, Mantelli MBH, Songo L: Thermal behavior analysis of wire mini heat pipe. J Heat Transf 2011, 133: 121502(1–9).

25. Meng KY, Jiang Z, Lian J, Jiang Q: Cu surfaces with controlled structure: from intrinsically hydrophilic to apparently superhydrophobic. Appl Surf Sci 2014, 290: 320–326.

26. Quéré D: Wetting and roughness. Annu Rev Mater Res 2008, 38: 71–99.

27. Zhang QY, Chen ZM, Qi ML: Study of hydrophilic films on copper tube. Adv Mater Res 2012, 581–582: 491–494.

28. Min JC, Wu XM, Shen LF, Gao F: Hydrophilic treatment and performance evaluation of copper finned tube evaporators. Appl Therm Eng 2011, 31: 2936–2942.

29. Min JC, Webb RL: Long-term wetting and corrosion characteristics of hot water treated aluminium and copper fin stocks. Int J Refrigeration 2002, 25(8):1054e1061.

30. Nam YS, Sharratt S, Cha G, Ju YS: Characterization and modeling of the heat transfer performance of nanostructured Cu micropost wicks. J Heat Transf 2011, 133: 101502(1–7).

31. Luo Y, Liu G, Zou LL, Yang YX, Wang XD: Experimental investigation of micro heat pipes for high-power light-emitting diode modules. Micro Nano Lett 2013, 8(10):646–649.

32. Wang XD, Zou LL, Liu G, Luo Y: Vacuum Fluid Charging and Packaging Technique for Micro Heat Pipes. 2013.

33. Luo Y, Liu G, Zou LL, Wang XD: Experimental investigation of micro silicon-Pyrex heat pipe for high-power LEDs. 2013.

10

CHAPTER

HEAT TRANSFER CHARACTERISTICS OF SQUARE MICRO PIN FINS UNDER NATURAL CONVECTION

Naoko Matsumoto, Toshio Tomimura, Yasushi Koito

Department of Advanced Mechanical Systems, Kumamoto University, Kumamoto, Japan

ABSTRACT

In order to comply with the recent demand for downsizing of the electric equipment, the minia- turization and the improvement in heat transfer performance of a heat sink under natural air-cooling are increasingly required. This paper describes the experimental and numerical investigations of heat sinks with miniature/micro pins and the effect of the pin size, pin height and the number of pins on heat transfer characteristics of heat sinks. Five types of basic heat sink models are investigated in this study. The whole heat transfer area of heat sinks having the different pin size, pin height and the number of pins respectively is kept constant. From a series of experiments and numerical analyses, it has been clarified that the heat sink temperature rises with increase in the number of pins. That is, the heat sink with miniaturized fine pins showed almost no effect on the heat transfer enhancement. This is because of the choking phenomenon occurred in the air space among the pin fins. Reflecting these results, it is confirmed that the heat transfer coefficient reduces with miniaturization of pins. Concerning the effects of the heat transfer area on the heat sink performance, almost the same tendency has been observed in other three series of large surface area, that is, higher pin height. Furthermore as a result of studying non-dimensional convection heat transfer performance, it was found that the relation between the Nusselt number (Nu) and the Rayleight number (Ra) is given by $Nu = 0.16\ Ra^{0.52}$.

Keywords: Natural Air-Cooling, Heat Sink, Micro Pin Fin, Heat Transfer Performance, Experiment, Numerical Analysis

1. INTRODUCTION

In electronic equipment applications with high heat fluxes, the heat transfer performance and the lifetime of components are often deteriorated due to temperature rise. To avoid those unfavorable problems, it is necessary to keep the component under a critical temperature by introducing a high performance heat sink. Furthermore, in order to comply with the recent demand for downsizing of the electronic equipment, the miniaturization and the improvement in heat transfer performance of the heat sink are required much more strongly than ever. But it is well known that the decrease in heat transfer area by downsizing leads to a decline of the heat sink performance.

Many studies of heat transfer characteristics using large scaled heat sinks have been carried out so far. For example, Aihara et al. [1] and Sara [2] revealed that the heat sink performance depends on the pin size, the population density of pin. Zografos and Sunderland [3] studied the effect of pin arrangement on the heat transfer performance, and reported that a choking phenomenon, which had a negative effect on heat transfer from a heat sink, was observed in the air space among pin fins. Huang et al. [4] , Sparrow et al. [5] and Sertkaya et al. [6] investigated experimentally the dependence of the pin fin performance on a heat sink orientation.

In electronic equipment applications, heat sinks are placed within enclosed areas and the emphasis is on making the enclosures more compact than ever. Yu et al. [7] and Bocu et al. [8] reported natural convection heat transfer from heat sink attached to enclosures in their studies.

On the other hand, concerning small sized heat sinks or nano particles structure, the analytical and experimental studies were published by Narasimhan et al. [9] and Minakami et al. [10] , Kunugi et al. [11] . Although interesting findings on natural and forced convection, and radiation from heat sinks have been reported in the previous studies, the mechanism of the heat transfer has not been clarified yet.

Accordingly, in the present study, the effects of pin size, pin height and the number of pins on the heat transfer characteristics of heat sinks with miniature/micro pins have been investigated fundamentally.

2. HEAT SINK MODELS

Five types of basic heat sink models are shown in Figure 1. The first model has four square pins, and is named as Type 1. Each pin placed at equal space of the heat sink base (length: 25 mm, width: 25 mm, thickness: 2 mm) has the pin width w_p = 6.25 mm and the pin height

h_p = 6.5, 14, 21.5, 29 mm at intervals of 7.5 mm. And in Type 2 model, the pin size is reduced to half the pin width and height of Type 1. That is, under the conditions of constant heat transfer area, the number of pins for Type 2 is quadrupled against Type 1. In the same manner, five types of heat sinks in total, which have the pin height from several hundred micro meters to a few millimetres and have the same heat transfer area, are prepared.

As listed in Table 1, the number of pins N is changed from 4 to 1024, and corresponding to this change in N, the pin width w_p and the pin height h_p vary from 0.39 mm to 6.25 mm and from 0.41 mm to 29 mm, respectively. The whole heat transfer area of the square pin series A_{hs} is 1.48 ´ 10^{-3} m^2, 2.23 ´ 10^{-3} m^2, 2.98 ´ 10^{-3} m^2 and 3.73 ´ 10^{-3} m^2.

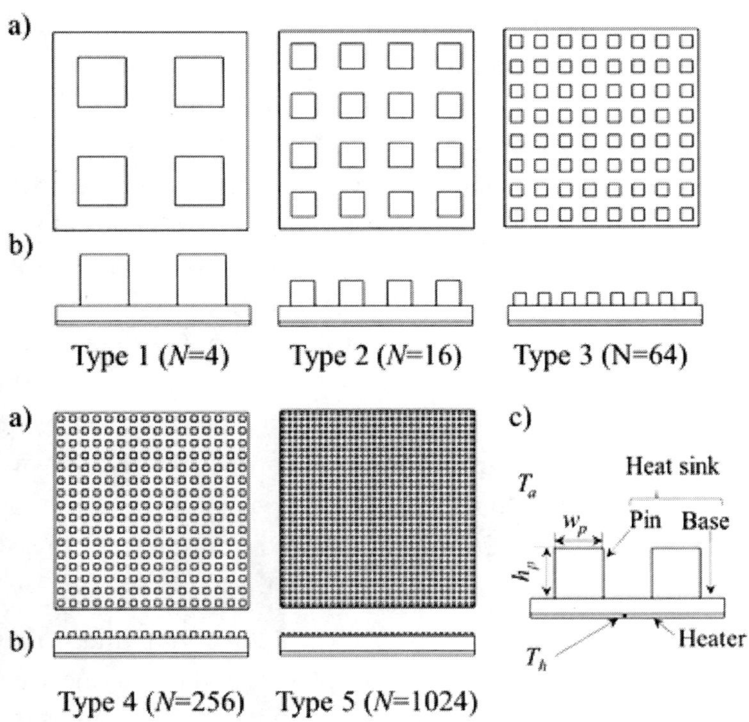

Figure 1. Heat sink models, (a) Top view, (b) Side view, (c) Parts and nomenclature.

Table 1. Dimensions of pin fin.

Type	N (–)	w_p (mm)	h_p (mm)			
			●	□	♦	Δ
1	4	6.25	6.5	14	21.5	29
2	16	3.13	3.25	7	10.75	14.5
3	64	1.56	1.63	3.5	5.38	7.25
4	256	0.78	0.81	1.75	2.69	3.63
5	1024	0.39	0.41	0.88	1.34	1.81

3. EXPERIMENTAL

Figure 2 shows a schematic diagram of experimental apparatus, which consists of a chamber, a vacuum pump, a vacuum gauge, and a heat sink with a heater used as the heat source. The chamber made of SUS is 400 mm in diameter and 170 mm in height. Here, its wall temperature which is regarded as the ambient temperature T_a was kept at 25°C by circulating cooling water. The heat sink was hung in the air in the chamber by using fine wires to reduce the heat loss from the heat sink to the chamber wall by conduction. The heat sink was placed in the upward facings of the heated surface. The pressure in the chamber was controlled with the vacuum pump from the atmospheric pressure to the vacuum condition of approximately 3.5×10^{-2} Pa, and the chamber pressure was measured using the vacuum gauge. Under the atmospheric pressure, heat is transferred from the heat sink to the surroundings by convection and thermal radiation. On the other hand, under the vacuum condition of approximately 3.5×10^{-2} Pa, heat is transferred from the heat sink to the surrounding wall only by thermal radiation.

As shown in Figure 3, the heat sink, which is made of Aluminum alloy A6063 (the thermal conductivity l = 200 W/(m·K), the emissivity e = 0.89) with a base plate of 25 mm long, 25 mm wide and 2 mm thick, was placed on the same sized square copper block (thickness: 2 mm, l = 398 W/(m·K)) through a thermal interface material (thickness:

0.5 mm, l = 3.1 W/(m·K)). Furthermore, on the bottom part of the copper block, a ceramic heater is attached through the thermal interface material. The representative temperature T_h was measured using a f 0.5 mm K-type thermocouple fixed in a small groove machined at the copper block center. In Figure 3, T_a is the ambient temperature.

First, under the atmospheric pressure, the heat rate Q_{in} = 1.25 W was applied to the heat sink by the ceramic heater. Second, the representative temperature T_h was measured under the steady-state condition. Third, under the vacuum condition of approximately 3.5×10^{-2} Pa, some amount of heat was applied to the heat sink by the heater until the heat sink temperature became the same temperature T_h under the atmospheric pressure, and then the heat transferred by thermal radiation Q_r from the heat sink under the steady-state conditions was measured based on the input power to the heater. Furthermore, in the same manner, T_h and Q_r for the heat sink consisting only of the base plate were also measured for reference.

4. NUMERICAL ANALYSIS

Figure 4 shows the boundary conditions for the large air space surrounding the heat sink placed in the upward facings of the heated surface. A series of numerical analyses have been performed for the physical models shown in Figure 1 and for the heat sink consisting only of the base plate (25 mm, 50 mm, 100 mm and 150 mm on a side) for reference. In this study, the fine wires for hanging the heat sink and the copper block, the thermal interface material were neglected for simplification, and only the heater (length: 25 mm, width: 25 mm, thickness: 0.5 mm), the heat sink, and the surrounding large air space were taken into consideration.

The present numerical calculations were carried out using the simulation software of Computational Fluid Dynamics (CFD). The following assumptions are applied in the analysis.

1) Fluid is incompressible.
2) Thermophysical properties are assumed to be constant except density (Boussinesq approximation).
3) Ambient temperature T_a is constant.

Based on these assumptions, the following basic equations are obtained under the steady state conditions.

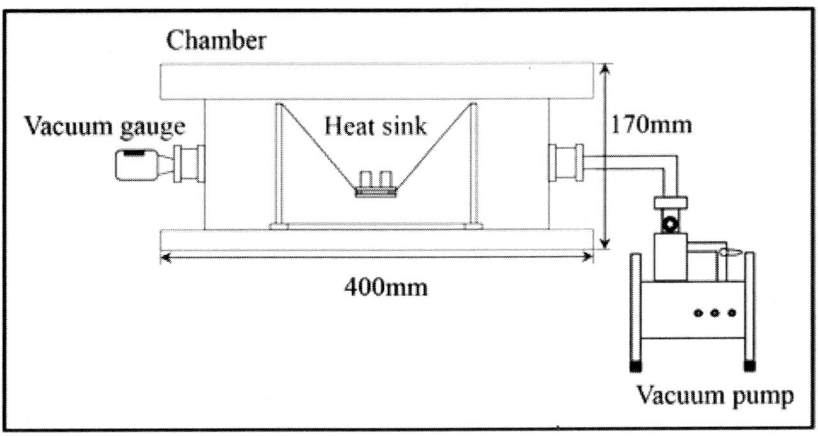

Figure 2. Schematic of experimental apparatus.

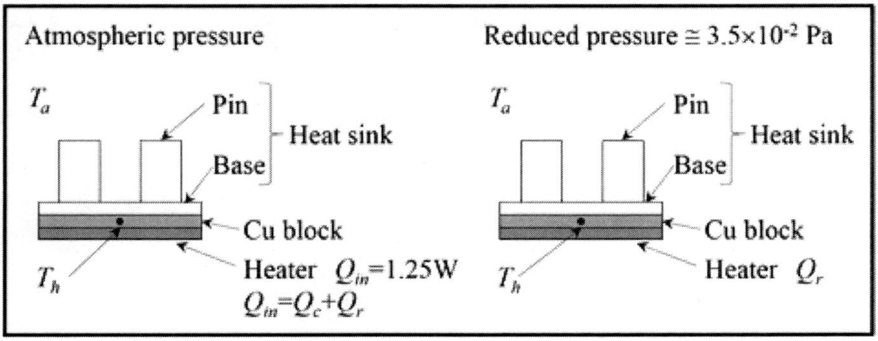

Figure 3. Configuration of tested heat sink.

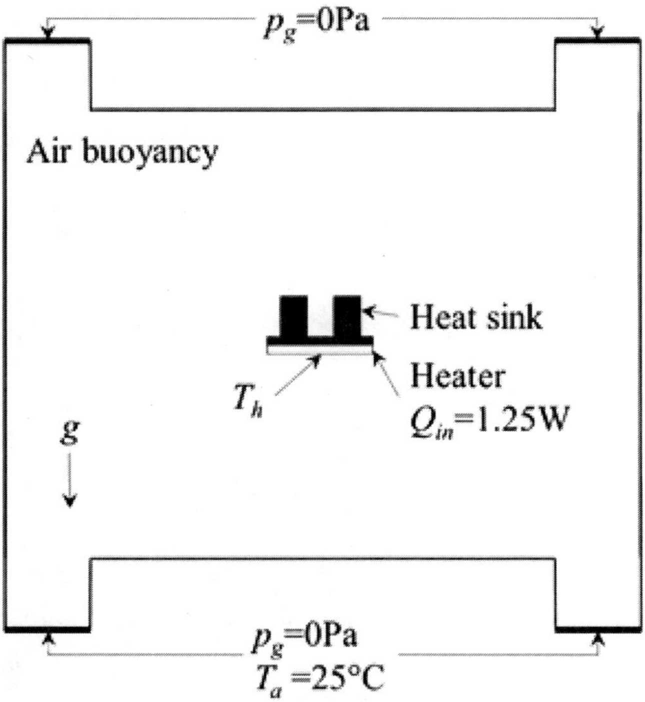

Figure 4. Boundary conditions of CFD (upward facing).

Continuity equation:

$$\frac{\partial u}{\partial x}+\frac{\partial v}{\partial y}+\frac{\partial w}{\partial z}=0 \quad (1)$$

Momentum equations:

$$u\frac{\partial u}{\partial x}+v\frac{\partial u}{\partial y}+w\frac{\partial u}{\partial z}=-\frac{1}{\rho}\frac{\partial p}{\partial x}+\upsilon\left(\frac{\partial^2 u}{\partial x^2}+\frac{\partial^2 u}{\partial y^2}+\frac{\partial^2 u}{\partial z^2}\right)$$

(x-direction) (2)

$$u\frac{\partial v}{\partial x}+v\frac{\partial v}{\partial y}+w\frac{\partial v}{\partial z}=-\frac{1}{\rho}\frac{\partial p}{\partial y}+\upsilon\left(\frac{\partial^2 v}{\partial x^2}+\frac{\partial^2 v}{\partial y^2}+\frac{\partial^2 v}{\partial z^2}\right)$$

(y-direction) (3)

$$u\frac{\partial w}{\partial x}+v\frac{\partial w}{\partial y}+w\frac{\partial w}{\partial z}=-\frac{1}{\rho}\frac{\partial p}{\partial z}+\upsilon\left(\frac{\partial^2 w}{\partial x^2}+\frac{\partial^2 w}{\partial y^2}+\frac{\partial^2 w}{\partial z^2}\right)+g\beta(T-T_a)$$

(z-direction) (4)

Energy equation:

$$u\frac{\partial T}{\partial x}+v\frac{\partial T}{\partial y}+w\frac{\partial T}{\partial z}=\alpha\left(\frac{\partial^2 T}{\partial x^2}+\frac{\partial^2 T}{\partial y^2}+\frac{\partial^2 T}{\partial z^2}\right) \quad (5)$$

Boundary conditions are given as follows. The same heat rate Q_{in} = 1.25 W as the experiment was applied to the heat sink. The velocity is zero on all the solid surfaces, that is u = v = w = 0. The gauge pressure p_g of the inlet and the outlet of the large surrounding space is 0 Pa, the inlet air temperature is 25°C, and other remaining surfaces are adiabatic. The amount of heat transfer rate by convection and thermal radiation is equal to that by conduction through the solid.

The equation of conduction within the solid is given by the following equation.

$$\frac{\partial^2 T}{\partial x^2}+\frac{\partial^2 T}{\partial y^2}+\frac{\partial^2 T}{\partial z^2}=0 \quad (6)$$

In the numerical analysis, the same heat rate Q_{in} = 1.25 W as the experiment was applied to the heater. In the case of large sized heat sink consisting only of the base plate, the heat rate added to the base plate was kept constant.

Then the representative temperature T_h of the heat sink under the steady-state conditions and the heat rate released from the heat sink by thermal radiation Q_r were calculated using the simulation software of CFD. In addition, the velocity vectors and the temperature fields at the central lateral plane of the heat sink were obtained.

5. RESULTS AND DISCUSSION

5.1. Heat Transfer Performance of Heat Sinks

Figure 5 and Figure 6 show the measured and simulated results on the effects of the number of pins N, the pin height h_p on the temperature rise DT of the heat sink. Here, DT is defined as the difference between the repre sentative temperature T_h and the ambient temperature T_a. Accordingly, the lower DT means a higher heat transfer performance. The symbols show the results for the heat sink with different pin height, that is, heat transfer area. The broken line shows the DT for the heat sink consisting only of the base plate. Further, the number of pins N corresponds to the heat sink type from Type 1 to Type 5.

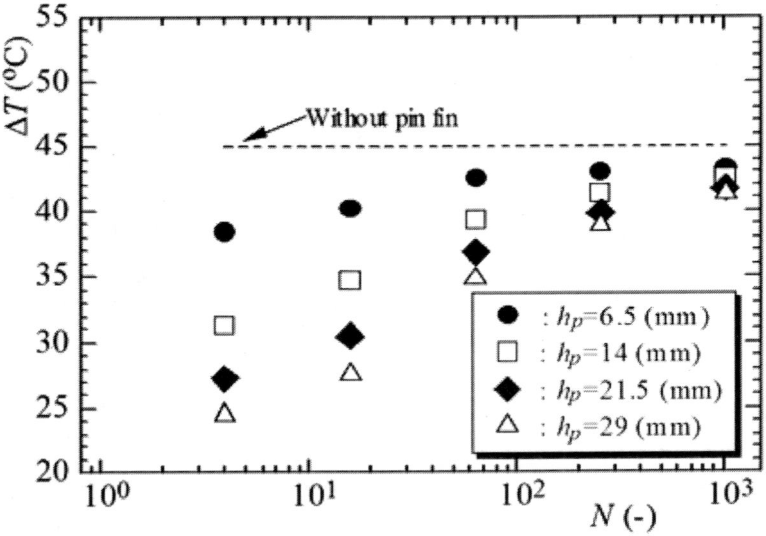

Figure 5. Effects of N and h_p on DT (measured results).

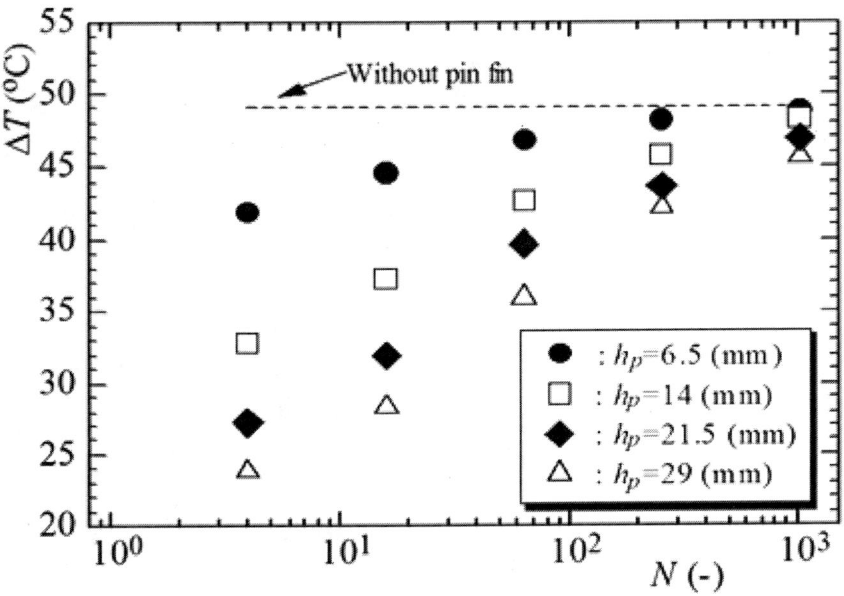

Figure 6. Effects of N and h_p on DT (calculated results).

Concerning Type 1 model having the smallest number of pins, it is found that the heat sink performance of Type 1 model is the highest and as the pin height h_p is higher, the heat sink performance is better. In a series of heat sinks, the heat transfer rate increases with the pin height of Type 1. Each representative temperature T_h of the heat sink rises with increase in the number of pins N. Especially, in the case of large N, the temperature rise DT of the heat sink comes close to that of the base plate level. In other words, the heat sink with miniaturized pins has almost no effect on the heat transfer enhancement. As seen from the figure, the simulated results agree qualitatively well with the measured results. That is, the dependence of the temperature rise DT on the number of pins N agrees fairly well with each other. Unfortunately, however, a quantitative agreement is not obtained between them. This is because the present physical model cannot reproduce the experimental heat sink system exactly.

5.2. Velocity Vector and Temperature Field

The velocity vector in air space surrounding the heat sink model at central longitudinal section, which are obtained from numerical simulation, are shown in Figure 7. From this figure, it can be seen that the velocity vector among pins becomes small with increase in the number of pins. In particular, the air velocity of Type 5 models becomes almost zero.

The temperature fields of the heat sink and its surrounding air space are shown in Figure 8. The representative temperature T_h at the heater surface is the highest. From this figure, it can be seen that the heater tem- perature rises from Type 1 to Type 5. Namely, this means that the increase in the number of pins causes the deterioration of the heat sink performance.

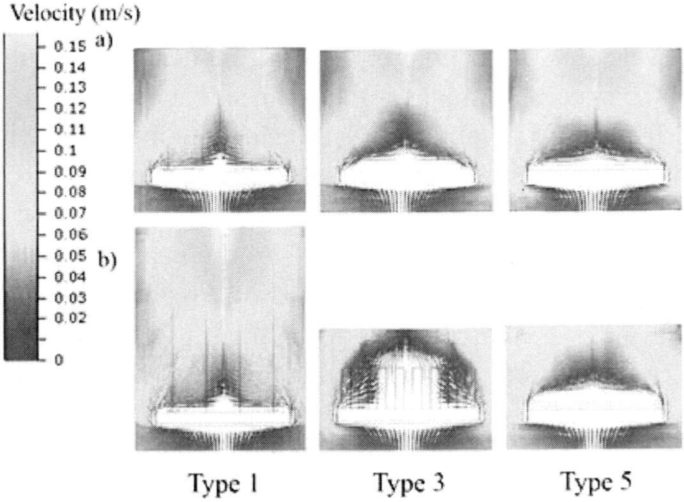

Figure 7. Flow field along center line of heat sink; (a) h_p = 6.5 mm, (b) h_p = 29 mm series.

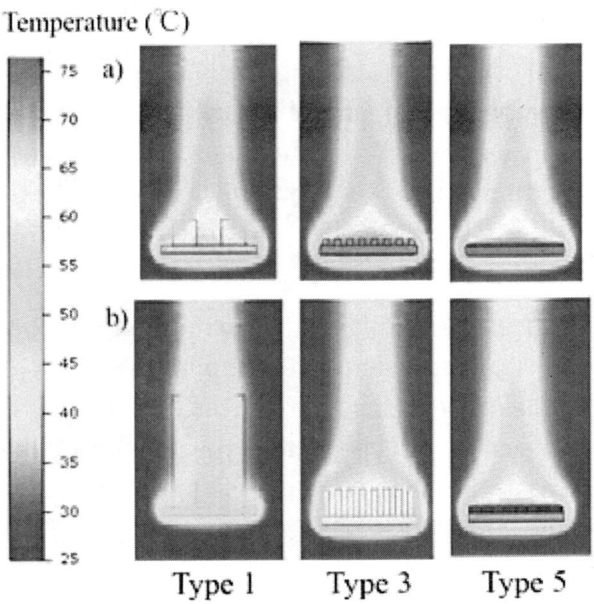

Figure 8. Temperature field along center line of heat sink; (a) h_p = 6.5 mm, (b) h_p = 29 mm series.

5.3. Air Velocity between Pins

Figure 9 shows the air velocity between pins which was numerically obtained for Type 1 to Type 5 models. The present air velocity means the average value obtained at central lateral plane of heat sink. As shown in the figure, the air velocity decreases with increase in the number of pins. This result shows the same tendency as the velocity vectors. In Type 3 to Type 5 models, regardless of the pin height, the air velocity becomes almost zero. This behavior must be the choking phenomena observed in the previous study using the large sized heat sinks done by Zografos et al. [3] , and as an inevitable consequence, it leads to low heat transfer performance of the heat sink.

5.4. The Ratio of Convective and Radiative Heat Transfer Rates

In order to estimate the heat sink performance from the view point of the heat transfer, the ratio of convective and radiative heat transfer rates in whole heat sink system has been also investigated experimentally and analytically, and the measured and calculated results are shown in Figure 10 and Figure 11 respectively. The total amount of heat transfer rate into and out has to keep the energy balance expressed by Equation (7). Under the conditions of the experiments and numerical calculations, Q_{in} is 1.25 W. In addition, the output energy from heat sink is the sum of convective and radiative heat transfer rates Q_c and Q_r, so the amount of heat transferred by convection Q_c is given by the following Equation (8).

$$Q_{\mathrm{in}} = Q_{\mathrm{out}} \tag{7}$$

$$Q_c = Q_{\mathrm{in}} - Q_r \tag{8}$$

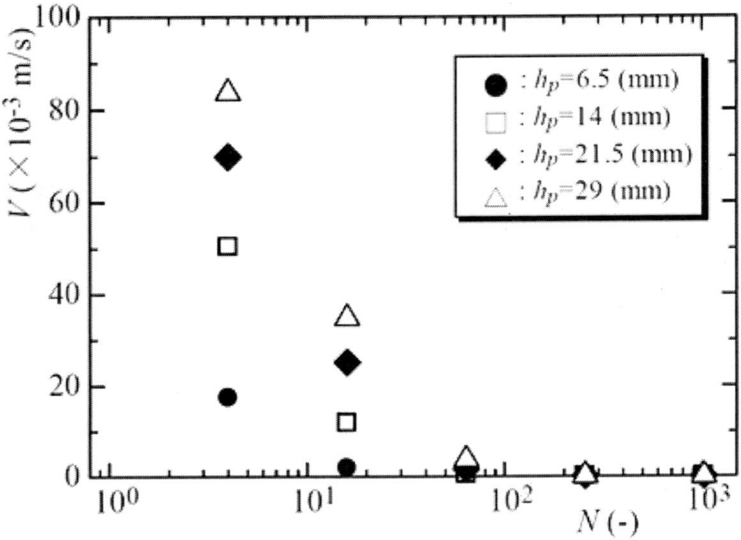

Figure 9. Flow velocity between pins on heat sink.

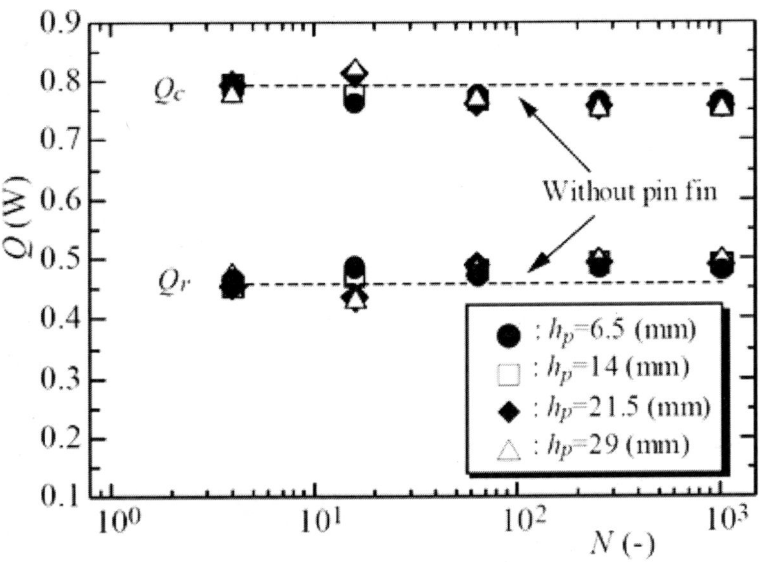

Figure 10. Effects of N and h_p on Q (measured results).

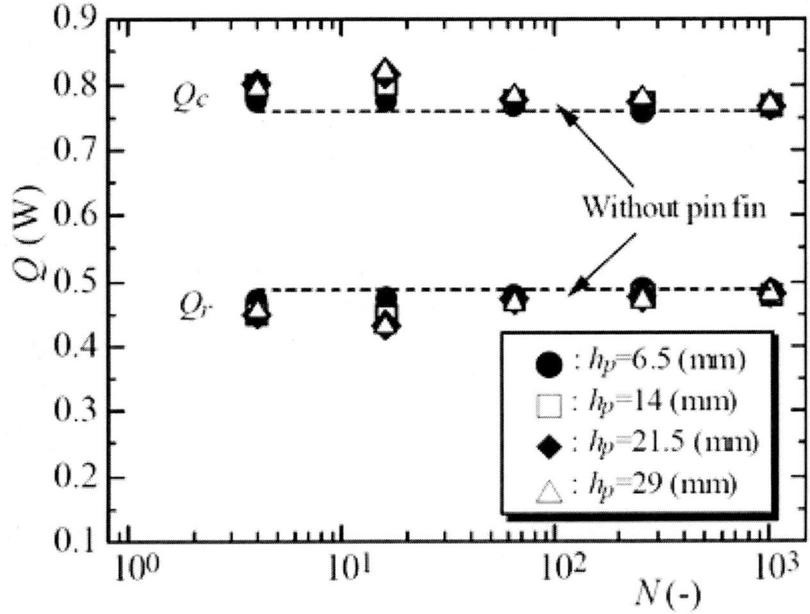

Figure 11. Effects of N and h_p on Q (calculated results).

As seen from Figure 10, it is clear that the convective component Q_c which is the dominant parameter in this case attains its maximum by Type 2, and reduces to Type 5 and then becomes saturated. It can be observed that the convection component Q_c of Type 2 is exactly proportional to the pin height h_p, in other words the heat rate transferred by thermal radiation Q_r vary inversely proportional to the pin height h_p. And in other heat sink models there is no correlation between the convection component Q_c and the pin height h_p. Furthermore, the heat rate transferred by convection Q_c occupies approximately 60% of the total input or output energy Q_{in} or Q_{out}, the heat rate transferred by thermal radiation Q_r is about 40%, and the result agrees very well with the results reported by Sparrow et al. [5] . And from Figure 11, it is found that the calculated results agree well with the measured results qualitatively. However a quantitative agreement is not obtained between them, as previously noted, this is because the present physical model cannot reproduce the experimental heat sink system exactly.

5.5. Heat Transfer of Heat Sink under Natural Convection

In numerical simulations, the total amount of heat transfer rates of the heater and the heat sink was investigated. The convective component of heat sink only is represented by Equation (9) using computable parameters of the total amount of convective heat transfer rate Q_c and the heat transfer from heater by convection Q_{c_ht}.

$$Q_{c\_hs} = Q_c - Q_{c\_ht} \tag{9}$$

The convective heat transfer rate of the heater is calculated by Equation (10),

$$Q_c = \frac{Nu\lambda_a}{L} A\left(T_h - T_a\right) \tag{10}$$

$$h = \frac{Nu\lambda_a}{L} \tag{11}$$

where Nu is the Nusselt number, l_a is thermal conductivity of air, L is characteristic dimension and A is heat transfer surface of the heater. Equation (11) gives the heat transfer coefficient h.

It is found that the heater has isothermal heating surface according to the simulated results, the Nusselt numbers at the lower and side surfaces of the heater are calculated by previously-proposed Equation (12) [12] , Equation (13) [13] and Equation (14), where Nu_1 is the Nusselt number of lower surface of heated plate, Nu_2 is that of vertical surface of heated plate. And Ra is the Rayleigh number which is represented as the correlation parameter between the Grashof number and the Prandtl number.

$$Nu_1 = 0.27Ra^{0.25} \tag{12}$$

$$Nu_2 = 0.59Ra^{0.25} \tag{13}$$

where

$$Ra = GrPr = \frac{g\beta(T_h - T_a)L^3}{\upsilon}\frac{\upsilon\rho c_p}{\lambda_a} \tag{14}$$

Two types of characteristic lengths L will be employed here. In the case of the lower surface of cooled plate, L is taken as L = A/P where A is the plate surface area and P is the perimeter which surrounds the area [14] . In the vertical flat plate, L is taken as 0.5 mm which is the height of heater. Using these numbers, Q_{c_ht} of the each side is calculated by substituting Nu_1 and Nu_2 into Equation (10). And then Q_{c_hs} is determined by Equation (9) using known Q_c.

Figure 12 shows the effects of the number of pins and the pin height on the convective heat transfer coefficient. Here, in order to evaluate the improvement in the convective heat transfer performance for heat sink base surface, the surface area of the top of the base is adopted as A (6.25×10^{-4} m^2 constant). In this figure, it is clear that each heat transfer coefficient h of the heat sink decreases with increase in the number of pins N and in the case of Type 5 model it comes close to that of the base plate level. Since Figure 12 agrees well with the results of the temperature rise, it is considered that the factor of decreasing heat transfer performance due to the miniaturization is the convective heat transfer coefficient.

Finally, the calculated data of heat sink, expressed in terms of Nu and Ra, are plotted in Figure 13. In this figure, the lower solid line represents Equation (15) [14] in which Nu_3 is the Nusselt number of upper surface of heated plate and the symbols (′) show the calculated results of the base plate.

$$Nu_3 = 0.54Ra^{0.25} \quad \left(10^4 < Ra < 10^7\right) \tag{15}$$

From these results, it is confirmed that Equation (15) applied to a range of the low-Rayleigh number. Using Equation (11), the Nusselt number of the heat sinks was calculated. Here, L was estimated to be the total value of the pin height h_p and the base height 2 mm. As seen this figure, regardless of the pin size, the pin height, the number of pins, it is revealed that Nu and Ra of the heat sink are expressed by an approximate equation, such as Equation (16).

$$Nu = 0.16Ra^{0.52} \tag{16}$$

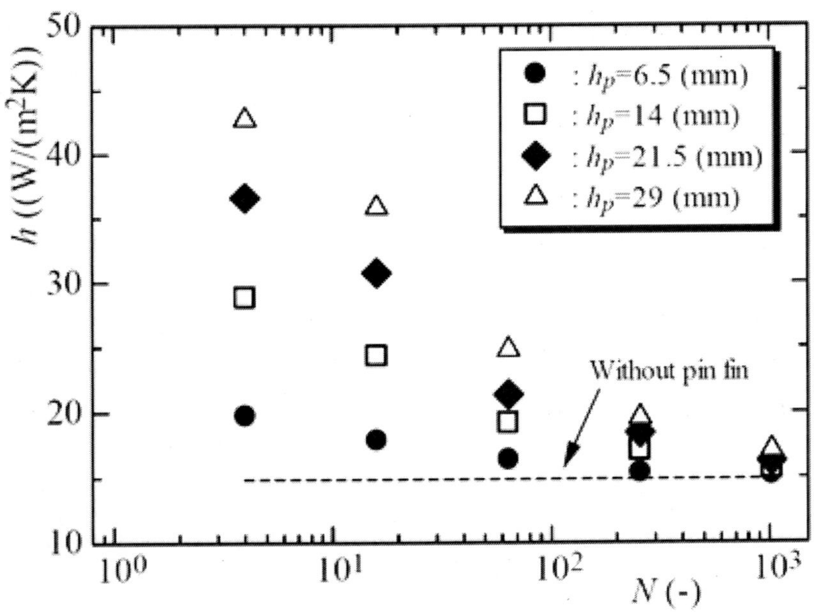

Figure 12. Effects of N and h_p on h (calculated results).

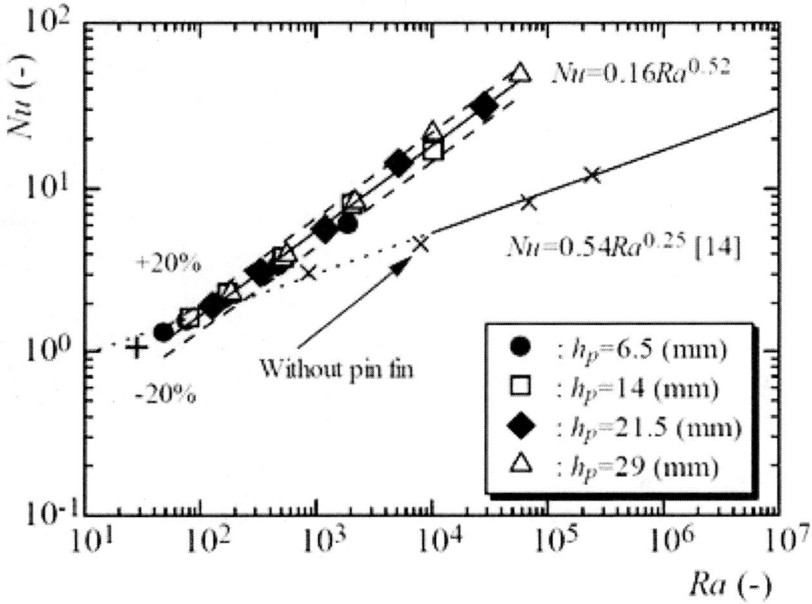

Figure 13. Correlation of Ra and Nu of the heat sink based on the base plate.

Since the calculated results were shown between ±20% of the approximate expression and with miniaturization of pins the Nusselt number Nu of the heat sink comes close to that of the base plate level, it is considered that the proposed correlation implies high accuracy.

6. CONCLUSIONS

In this study, experimental and numerical studies have been done on the effects of the pin size, pin height, the number of pins on the heat transfer performance of the heat sink.

Concerning the heat sinks with the same heat transfer area, it has been confirmed that the heat sink performance changes depending on the population density of pin and the pin size. From the measured and calculated results, the heat sink temperature has been shown to rise with increase in the number of pins. Especially, the heat sink with miniaturized pins has almost no effect on the heat transfer enhancement. It is considered that these characteristics of the heat sink cause the decrease in the convective component, which occupies approximately 60% of total heat transfer rate, and the heat transfer coefficient due to the decrease in the flow velocity when the pin to pin clearances are reduced.

Regardless of the pin size, the pin height, and the number of pins, it has been revealed that the Nusselt number and the Reyleigh number of the heat sink are represented by an approximate equation.

REFERENCES

1. Aihara, T., Maruyama, S. and Kobayakawa, S. (1990) Free Convective/Radiative Heat Transfer from Pin Fin Arrays with a Vertical Base Plate (General Representation of Heat Transfer Performance). International Journal of Heat Mass Transfer, 33, 1223-1232.

2. Sara, O.N. (2003) Performance Analysis of Rectangular Ducts with Staggered Square Pin Fins. Energy Conversion and Management, 44, 1787-1803.

3. Zografos, A.I. and Sunderland, J.E. (1990) Natural Convection from Pin Fin Arrays. Experimental Thermal and Fluid Science, 3, 440-449.

4. Huang, R.T., Sheu, W.J. and Wang, C.C. (2008) Orientation Effect on Natural Convective Performance of Square Pin Fin

Heat Sinks. International Journal of Heat and Mass Transfer, 51, 2368-2376.

5. Sparrow, E.M. and Vemuri, S.B. (1986) Orientation Effects on Natural Convection/Radiation Heat Transfer from Pin- Fin Arrays. International Journal of Heat and Mass Transfer, 29, 359-368.

6. Sertkaya, A.A., Bilir, S. and Kargici, S. (2011) Experimental Investigation of the Effects of Orientation Angle on Heat Transfer Performance of Pin-Finned Surfaces in Natural Convection. Energy, 36, 1513-1517.

7. Yu, E. and Joshi, Y. (2002) Heat Transfer Enhancement from Enclosed Discrete Components Using Pin-Fin Heat Sinks. International Journal of Heat and Mass Transfer, 45, 4957-4966.

8. Bocu, Z. and Altac, Z. (2011) Laminar Natural Convection Heat Transfer and Air Flow in Three-Dimensional Rectangular Enclosures with Pin Arrays Attached to Hot Wall. Applied Thermal Engineering, 31, 3189-3195.

9. Narasimhan, S. and Majdalani, J. (2002) Characterization of Compact Heat Sink Models in Natural Convection. IEEE Transactions on Components and Packaging Technologies, 25, 78-87.

10. Minakami, K., Mochisuki, S., Murata, A., Yagi, Y. and Iwasaki, H. (1993) Heat Transfer Characteristics of Pin-Fins with In-Line Arrangement (1st Report, Effect of the Pin Pitch). Japan Society of Mechanical Engineers, 59, 300-307.

11. Kunugi, T., Muko, K. and Shibahara, M. (2004) Ultrahigh Heat Transfer Enhancement Using Nano-Porous Layer. Superlattices and Microstructures, 35, 531-542.

12. Kreith, F. and Bohn, M.S. (2000) Principles of Heat Transfer. 6th Edition, Brooks/Cole.

13. Nakayama, A. (2008) Problems in Heat Transfer. Japan Society of Mechanical Engineers. Maruzen Publishing Co., Ltd.

14. Goldstein, R.J. and Sparrow, E.M. (1973) Natural Convection Mass Transfer Adjacent to Horizontal Plates. International Journal of Heat and Mass Transfer, 16, 1025-1035.

11

CHAPTER

CFD SIMULATION OF THE EFFECT OF PARTICLE SIZE ON THE NANOFLUIDS CONVECTIVE HEAT TRANSFER IN THE DEVELOPED REGION IN A CIRCULAR TUB

Reza Davarnejad[1]*, Sara Barati[1] and Maryam Kooshki[2]

(1)Department of Chemical Engineering, Faculty of Engineering, Arak University

(2)Department of Chemistry, Faculty of Science, Arak University

ABSTRACT

The CFD simulation of heat transfer characteristics of a nanofluid in a circular tube under constant heat flux was considered using Fluent software (version 6.3.26) in the laminar flow. Al_2O_3 nanoparticles in water with concentrations of 0.5%, 1.0%, 1.5%, 2% and 2.5% were used in this simulation. All of the thermo-physical properties of nanofluids were assumed to be temperature independent. Two particle sizes with average size of 20 and 50 nm were used in this research. It was concluded that heat transfer coefficient increased by increasing the Reynolds number and the concentration of nanoparticles. The maximum convective heat transfer coefficient was observed at the highest concentration of nano-particles in water (2.5%). Furthermore, the two nanofluids showed higher heat transfer than the base fluid (water) although the nanofluid with particles size of 20 nm had the highest heat transfer coefficient.

Keywords; Nanofluid Heat transfer Simulation Laminar flow

INTRODUCTION

Fluid heating and cooling play the significant roles in a lot of industrial processes such as power stations, production processes, transportation and electronics. Most of the methods for heat transfer are based on the structure variation, vibration of the heated surface, injection or suction of fluid and applying electrical or magnetic fields (Ahuja 1975; Bergles & Webb 1970; Bergles 1973). These techniques meet a great enhancement in heat flux. Heat transfer in the traditional fluids such as water, ethyleneglycol and oil inherently has low thermal conductivity compared with the metals and metal oxides. Therefore, fluids with suspended solid particles are expected to have better heat transfer properties (Eastman et al. 2001). Due to the associated technological problems, the majority of studies on heat transfer of suspension of metal oxides in fluids were limited to suspensions with millimeter or micron-sized particles. The large particles may cause severe problems in the heat transfer equipments. In particular, large particles quickly tend to settle out. So, pressure drop can occur in the micro-channels (Khanafer et al. 2003). Furthermore, the abrasive actions of the particles cause the erosion phenomenon in pipe lines. Small particles and their little volume fractions prevent particles clogging and pressure drop increment in the nanofluids (Khanafer et al. 2003; Zhou 2004). Moreover, large surface area of nanoparticles increases the stability and reduces the sedimentation of nanoparticles. A more dramatic improvement in heat transfer efficiency is expected as a result of the particle size reduction in a suspension because heat transfer takes place at the particles surface (ZeinaliHeris et al. 2007).

Choi employed the particles in nanometer dimensions as a suspended solution (Choi & Eastman 1995). He showed that the nanofluid thermal conductivity considerably increased. Lee et al. showed that the suspension of 4.0% with 35 nm Cuo particles in ethylene glycol had 20% increment in the thermal conductivity (Lee et al. 1999). Choi et al. observed 60% enhancement in the thermal conductivity of engine oil with 1.0% carbon nanotube (Choi et al. 2001). Das et al. investigated the temperature dependency of thermal conductivity in the nanofluids (Das et al. 2003). It was observed that a 2–4 fold increase in the thermal conductivity of nanofluid can take place over a temperature range of 21–51°C.

According to the literature, alumina and copper oxide are the most ordinary and cheap nanoparticles which are used in the applied processes (Rezaee & Tayebi 2010). Xuan and Li experimentally studied the convective heat transfer and friction coefficient for the nanofluid in both laminar and turbulent flows (Xuan & Li 2000; Xuan & Li 2003). According to these researches, the flow velocity and volume fraction of nanoparticles affected the heat transfer coefficient. Wen and Ding investigated the convective heat transfer characteristics in Al_2O_3-water nanofluid along a tube (Wen & Ding 2004). It was observed that heat transfer increased by increasing the Reynolds number and volumetric ratio of particles.

Abu-Nada investigated the effects of variable viscosity and thermal conductivity of a nanofluid (Al_2O_3-water) on the natural convective heat transfer (Abu-Nada 2009). Sharma et al. experimentally studied the convective heat transfer coefficient and pressure drop in the transient region for Al_2O_3-water nanofluid under a constant heat flux

(Sharma et al. 2009). They found that convective heat transfer increased by adding Al_2O_3 nanoparticles in water.

Mirmasoumi et al. numerically studied the convective heat transfer in a fully developed flow for Al_2O_3-water nanofluid (Mirmasoumi & Behzadmehr 2008). They applied two-phase mixture model in their simulation. They found that the convective heat transfer coefficient significantly increased by decreasing the nanoparticles mean diameter. Since the theoretical models such as Maxwell and Hamilton-Crosser (Keblinski et al. 2002; Xue & Xu 2005; Eastman et al. 2001) predict the thermal conductivity of nanofluids, the mechanisms of thermal conductivity enhancement in the nanofluids should be studied.

Moraveji et al. simulated water-Al_2O_3 nanofluid through a tube under constant heat flux (Moraveji et al. 2011). They found that the heat transfer coefficient increased by increasing the nanoparticle concentration and Reynolds number. Furthermore, the heat transfer coefficient increased by decreasing particle diameter.

In this research, the convective heat transfer in the developed region of the tube flow containing water and Al_2O_3 nanofluid under constant heat flux was simulated using the Computational Fluid Dynamics (CFD) tools. Al_2O_3 nanoparticles with two average diameters of 20 nm and 50 nm were dispersed in water. The nanofluids with five different Al_2O_3 nanoparticle concentrations (0.2%, 1.0%, 1.5%, 2.0% and 2.5% volume fraction) were used. Effects of nanoparticle size and their concentrations on the convective heat transfer coefficient were also investigated.

MATHEMATICAL MODELING

The nanofluid as a single phase fluid with different physical properties (density, thermal conductivity and viscosity) was applied for the simulation. Heat transfer and flow are considered by the continuity, momentum and energy equations (Izadi et al. 2009):

Continuity equation:

$$\nabla.\left(\rho_{nf} V_m\right) = 0 \qquad (1)$$

Momentum equation:

$$\nabla.\left(\rho_{nf} V_m V_m\right) = -\nabla P + \nabla.\left(\mu_{nf} \nabla V_m\right) \qquad (2)$$

Energy equation:

$$\nabla\left(\rho_{nf} V_m T C\right) = \nabla\left(k_{nf} \nabla T\right) \qquad (3)$$

The physical properties for above equation can be obtained (Ghasemi & Aminossadati 2009):

$$\rho_{nf} = (1-\phi)\rho_{bf} + \phi\rho_p \qquad (4)$$

The effective heat capacity is calculated by (Ghasemi & Aminossadati 2011):

$$C_{nf} = \frac{\phi(\rho C)_p + (1-\phi)(\rho C)_{bf}}{\rho_{nf}} \qquad (5)$$

The viscosities of nanofluid (with the average particle size of 20 and 50 nm) can be predicted by Einstein's equation:

$$\mu_{nf} = \mu_{bf}(1 + 2.5\phi) \qquad (6)$$

In this research, the single-phase approach was applied. Solid particles with less than 100 nm diameter were spotted in the single-phase approach. Moreover, some necessary data for the thermal conductivities determination for various concentrations of nanofluid were obtained from the literature (ZeinaliHeris et al. 2007). Yu and Choi's correlation (Yu & Choi 2003; Trisaksri & Wongwises 2007) was applied for the nanofluid effective thermal conductivity determination:

$$k_{nf} = \left[\frac{k_p + 2k_{bf} + 2(k_p - k_{bf})(1+\beta)^3\phi}{k_p + 2k_{bf} - 2(k_p - k_{bf})(1+\beta)^3\phi}\right]k_{bf} \tag{7}$$

Where, β is the ratio of the nano-layer thickness to the original particle radius (β = 0.1) (Yu & Choi 2003). The rheological and physical properties of the nanofluid were calculated at the average bulk temperature.

As shown in Figure 1, a two dimensional pipe (with 1 m length and 6 mm inner diameter) was spotted in our simulation. The single phase approach was used for nanofluid simulation and effect of nanoparticle concentration on the convective heat transfer coefficient was investigated in the various Reynolds numbers (700 < Re < 2050).There were 20 meshes in the radial direction with a size ratio of 1 from the center to the wall. The constant heat flux of 1.8 (W/cm^2) as a boundary condition at the pipe wall was applied.

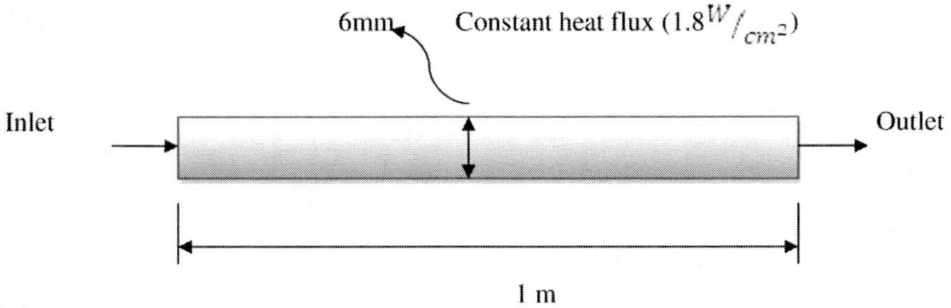

Figure 1. Pipe numerical domain.

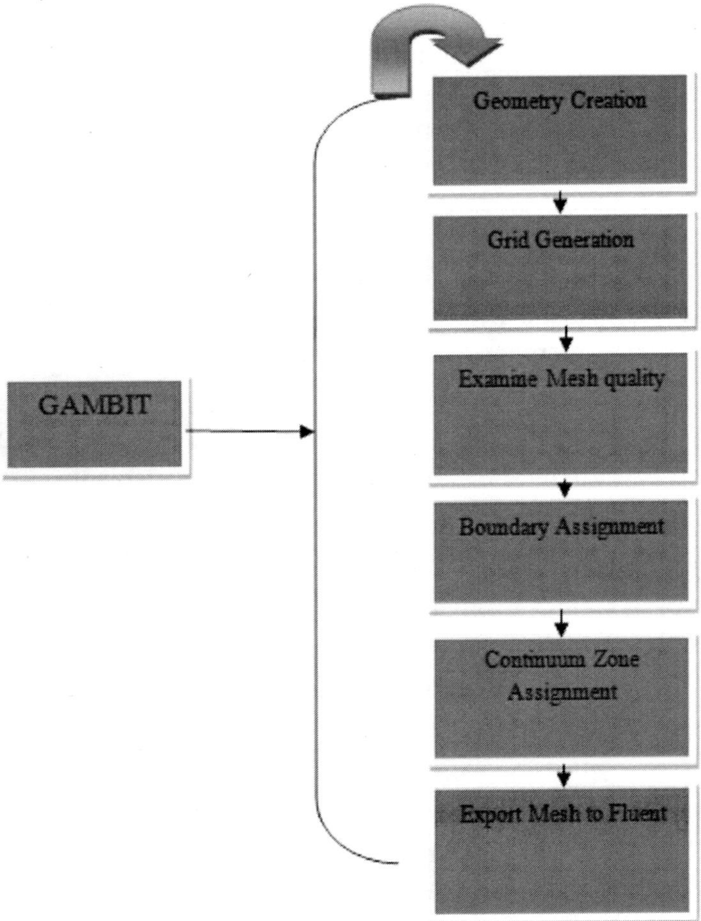

Figure 2. Temperature map at the end part of the pipe for the nanofluid (20 NM & 2.5%). **(A).** GAMBIT steps during meshes generation. (**B**). Meshes generation. (**C**). U = 0.1664 (m/s). (**D**). U = 0.0706 (m/s).

CFD simulation procedure

The geometry and the gird were generated using GAMBIT. The GAMBIT is an integrated preprocessor for CFD analysis. The sequences of GAMBIT steps are shown in Figure 2a. There are 20 meshes in the radial direction with a size ratio of 1 from the center to the wall. Further, there are 1000 meshes in the horizontal direction with an average size. The sequences of GAMBIT steps are shown in Figure 2a. The physical boundary conditions for the geometry are defined as inlet, outlet and wall of the pipe. The continuum was the fluid. Then, the mesh file was successfully conducted into the Fluent (version 6.3.26). For single phase approach, solid particles with diameter less than 100 nm were spotted. Therefore, single phase approach was adopted for nanofluid modeling (Moraveji et al. 2011). The fluid was entered the pipe with a constant velocity in each run. The initial temperature of fluid (at t = 0) was 25°C. The Symmetric option was chosen in the software. The constant heat flux of 1.8 (W/cm^2) as a boundary condition at the pipe wall was applied.

RESULTS AND DISCUSSION

The local heat transfer coefficient and local Nusselt number were calculated using the following equations:

$$Nu(x) = \frac{h(x)D}{k} \tag{8}$$

$$h_{nf}(x) = \frac{q}{(T_w(x) - T_f(x))} \tag{9}$$

Where, D, q, h_{nf}, k, T_w and T_{nf} are pipe diameter, heat flux, nanofluid heat transfer coefficient, thermal conductivity of the fluid, tube wall temperature and nanofluid temperature, respectively.

Figure 2b demonstrates the meshes generation in this research. Figure 2c and d show temperature distribution at the end of pipe for

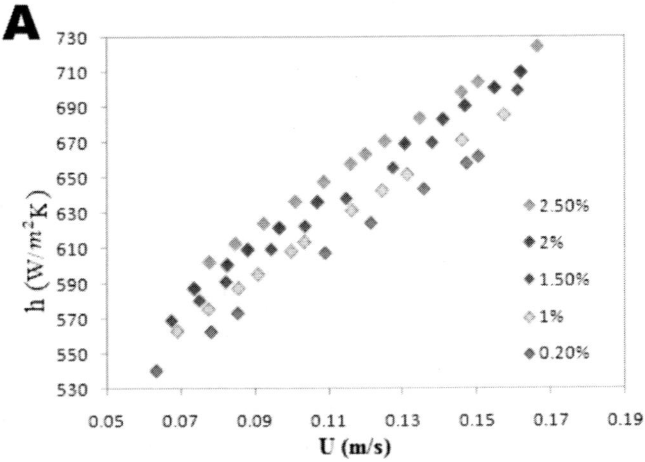

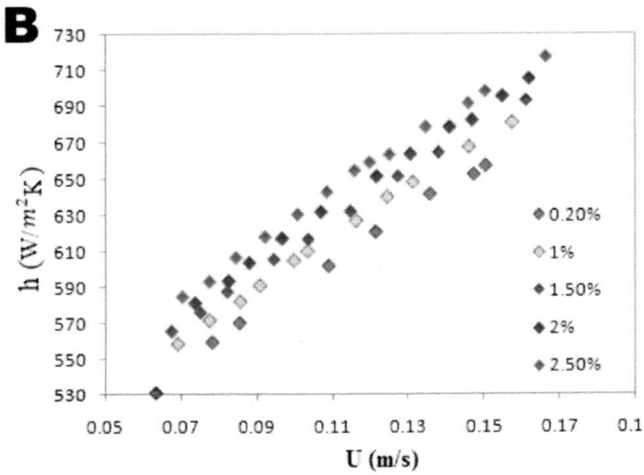

(a) Nano particle 50 nm

Figure 3. (A). Heat transfer coefficient versus velocity for the nanofluid with particle size of 20 nm. **(b)**. Heat transfer coefficient velocity for the nanofluid with particle size of 50 nm.

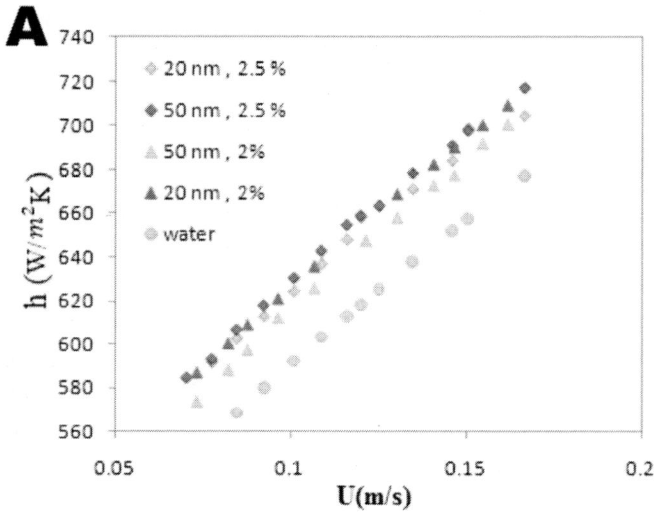

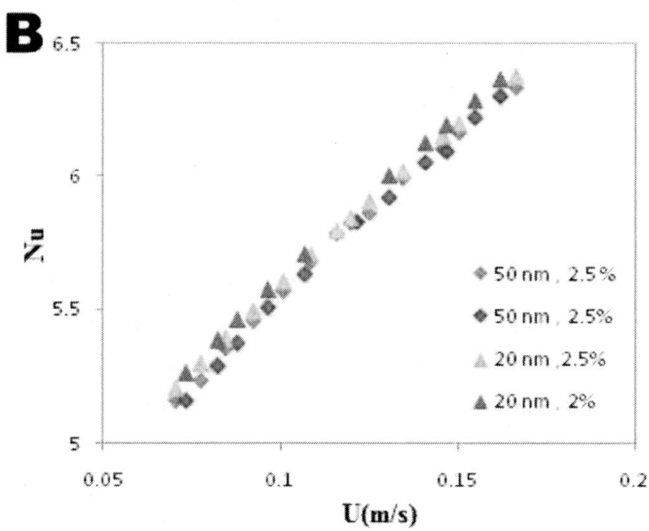

Figure 4. (A) . Effect of particle size on the heat transfer coefficient in the various velocities. (**b**). Effect of particle size on the Nusselt number in the various velocities.

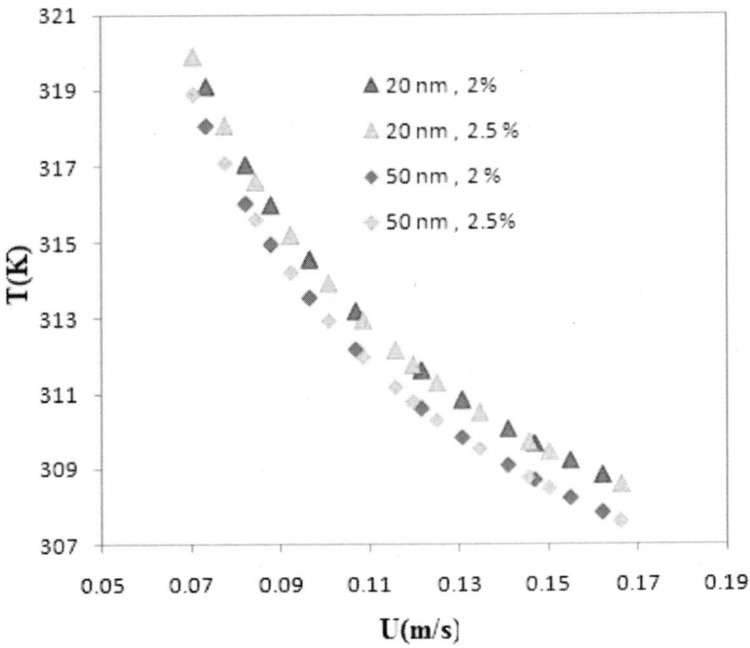

Figure 5. Temperature of nanofluid versus velocity for various concentrations of nanoparticles.

velocity of 0.0706 and0.1664 (m/s) (or Re = 420 and 991.6) for nanofluid (with the particle diameter of 20 nm and volume fraction of 2.5). The nanofluid was heated by the pipe wall and its temperature increased along the pipe. The temperature of nanofluid along the pipe at Reynolds number of 420 varied sharper than that of Reynolds number of 991.6.Its reason was due to magnifying the heat transfer coefficient (He et al. 2009).

Figure 3a and b show heat transfer coefficient in the tube versus velocity for various concentrations of nanoparticles (with diameters of 20 and 50 nm). According to these figures, heat transfer coefficients increased by increasing the volume fraction and velocity.

Figure 4a and b show heat transfer coefficients (h) and Nusselt (Nu) number in the tube versus velocity at particle concentrations of 2.5

and 2% for both particle sizes. As shown in these figures, heat transfer coefficients increased by increasing velocity and decreasing particle diameter. Furthermore, the nanofluid with average particle size of 20 nm showed the maximum heat transfer (Anoop et al. 2009).

Figure 5 shows nanofluid temperature versus velocity for particles of 20 and 50 nm (for volume fractions of 2.5 and 2). As show in this figure, the nanofluid temperature decreased by increasing the particle size although a change in the volume fraction had no significant effect on the nanofluid temperature. The experimental results investigated by Anoop et al. (2009) were in very good agreement with our theoretical data and supported them.

CONCLUSIONS

In this article, the heat transfer coefficient in the developed region of pipe flow containing Al_2O_3-water nanofluid during the constant heat flux was simulated using CFD. The focal point of investigation was to evaluate the effect of particle size on convective heat transfer characteristics in the developed region of the tube flow containing water-Al_2O_3 nanofluid. It was observed that both nanofluids (with 20 and 50 nm particles size) showed higher heat transfer characteristics than that of the base fluid (water). Furthermore, the nanofluid with 20 nm particles size showed the highest heat transfer coefficient. The average heat transfer coefficient and Nusselt number increased by increasing the particle concentration and flow rate. The average temperature of nanofluid decreased by increasing the particles size.

NOMENCLATURE

D diameter of the tube [m]

k thermal conductivity [W/m.K]

Nu Nusselt number

Re Reynolds number

T Temperature [K]

C specific heat $(J/_{kg.K})$

q heat flux [W/]

x distance a long axis [m]

Subscripts

w wall

nf nanofluid

In inlet

p particle

Out outlet

bf base fluid

f fluid

Greek letters

μ viscosity (Pa.s)

ϕ volume fraction

AUTHORS' CONTRIBUTIONS

All of the authors, Dr. RD (from Chemical Eng. Dep. of Arak University), Ms. SB (MSc) (from Chemical Eng. Dep. of Arak University) and Ms. MK (MSc) (from Chemistry Dep. of Arak University), read the GP version of article, carefully and approved it for publication in the SpringerPlus.

REFERENCES

1. Abu-Nada E: Effects of variable viscosity and thermal conductivity of Al2O3–water nanofluid on heat transfer enhancement in natural convection. Int J Heat Fluid Flow 2009, 30: 679-690.
2. Ahuja AS: Augmentation of heat transport in laminar flow of polystyrene suspension, I. Experimental and results. J Appl Phys 1975, 46(8):3408-3416.
3. Anoop KB, Sundararajan T, Das SK: Effect of particle size on the convective heat transfer in nanofluid in the developing region. Int J Heat Mass Transf 2009, 52: 2189-2195.
4. Bergles AE: Recent development in convective heat transfer augmentation. Appl Mech Rev 1973, 26: 675-682.
5. Bergles AE, Webb RL: Augmentation of convective heat and mass transfer. ASME, New York, N.Y.: Winter Annual Meeting of the American Society of Mechanical Engineers; 1970.
6. Choi SUS, Eastman JA: Enhancing thermal conductivity of fluid with nanoparticles. San Francisco, CA: ASME International Mechanical Engineering Congress & Exhibition; 1995.

7. Choi SUS, Zhang ZG, Yu W, Lockwood FE, Grulke EA: Anomalous thermal conductivity enhancement in nanotube suspensions. Appl Phys Lett 2001, 79(14):2252-2254.

8. Das SK, Putra N, Theisen P, Roetzel W: Temperature dependence of thermal conductivity enhancement for nanofluid. Journal of Heat Transfer 2003, 125: 567-574.

9. Eastman JA, Choi SUS, Li S, Yu W, Thompson LJ: Anomalously increased effective thermal conductivities of ethylene glycol-based nanofluids containing copper nanoparticles. Appl Phys Lett 2001, 78(6):718-720.

10. Ghasemi B, Aminossadati SM: Natural convection heat transfer in an inclined enclosure filled with a water-CuO nanofluid. Numerical Heat Transfer, Part A 2009, 55: 807-823.

11. Ghasemi B, Aminossadati SM: A numerical study on the forced convection of laminar nanofluid in a micro channel with both slip and no slip conditions. Numerical Heat Transfer, Part A 2011, 59: 114-129.

12. He Y, Men Y, Zhao Y, Lu H, Ding Y: Numerical investigation into the convective heat transfer of TiO2nanofluids flowing through a straight tube under the laminar flow conditions. Appl Therm Eng 2009, 29: 1965-1972.

13. Izadi M, Behzadmehr A, Jalali-Vahid D: Numerical study of developing laminar forced convection of a nanofluid in an annulus. Int J Therm Sci 2009, 48: 2119-2129.

14. Keblinski P, Phillpot SR, Choi SUS, Eastman JA: Mechanism of heat flow in suspension of nano-sized particle (nanofluids). Int J Heat Mass Transf 2002, 45(4):855-863.

15. Khanafer K, Vafai K, Lightstone M: Buoyancy-driven heat transfer enhancement in a two-dimensional enclosure

utilizing nanofluids. Int J Heat Mass Transf 2003, 46(19):3639-3653.

16. Lee S, Choi SUS, Li S, Eastman JA: Measuring thermal conductivity of fluids containing oxide nanoparticles. Journal of Heat Transfer 1999, 121: 280-289.

17. Mirmasoumi S, Behzadmehr A: Effect of nanoparticles mean diameter on mixed convection heat transfer of a nanofluid in a horizontal tube. Int J Heat Fluid Flow 2008, 29: 557-566.

18. Moraveji MK, Darabi M, Hossein Haddad SM, Davarnejad R: Modeling of convective heat transfer of a nanofluid in the developing region of tube flow with computational fluid dynamics. Int Commun Heat Mass Transfer 2011, 38: 1291-1295.

19. Rezaee FK, Tayebi A: Energy destruction of forced convective (EG + AL2O3) nanofluid through a duct with constant wall temperature in contrast to (Ethylene Glycol) fluid. J Appl Sci 2010, 10(13):1279-1285.

20. Sharma KV, SyamSundar L, Sarma PK: Estimation of heat transfer coefficient and friction factor in the transition flow with low volume concentration of Al2O3nanofluid flowing in a circular tube and with twisted tape insert. Int Commun Heat Mass Transfer 2009, 36: 503-507.

21. Trisaksri V, Wongwises S: Critical review of heat transfer characteristics of nanofluids. Renew Sustain Energy Rev 2007, 11: 512-523.

22. Wen D, Ding Y: Experimental investigation into convective heat transfer of nanofluids at the entrance region under laminar flow conditions. Int J Heat Mass Transf 2004, 47: 5181-5188.

23. Xuan YM, Li Q: Heat transfer enhancement of nanofluids. Int J Heat Fluid Flow 2000, 21: 58-64.

24. Xuan YM, Li Q: Investigation on convective heat transfer and flow features of nanofluids. Journal of Heat Transfer 2003, 125: 151-155.

25. Xue Q, Xu WM: A model of thermal conductivity of nanofluids with interfacial shells. Mater Chem Phys 2005, 90(2):298-301.

26. Yu W, Choi SUS: The role of international layers in the enhanced thermal conductivity of nanofluids: A renovated Maxwell model. J Nanoparticle Res 2003, 5: 167-171.

27. ZeinaliHeris S, Nasr Esfahany M, Etemad SG: Experimental investigation of convective heat transfer of Al2O3/water nanofluid in circular tube. Int J Heat Fluid Flow 2007, 28: 203-210.

28. Zhou DW: Heat transfer enhancement of copper nanofluid with acoustic cavitations. Int J Heat Mass Transf 2004, 47(14):3109-3117.

12
CHAPTER

CALCULATION METHOD FOR FORCED-AIR CONVECTION COOLING HEAT TRANSFER COEFFICIENT OF MULTIPLE ROWS OF MEMORY CARDS

Takayuki Atarashi[1,2*], Tetsuya Tanaka[1], Shigeyasu Tsubaki[3], Shigeki Hirasawa[2]

[1]Hitachi, Ltd., Infrastructure Systems Company, Toshima, Tokyo, Japan

[2]Department of Mechanical Engineering, Kobe University, Kobe, Japan

[3]Hitachi, Ltd., Information and Telecommunication Systems Company, Hatano, Japan

ABSTRACT

Forced-air convection cooling of high-power electronic devices is widely used, but it has a problem that a rise in temperature of the air used to cool the upstream devices decreases the cooling capability for the downstream devices. In this study we made an experimental apparatus including a memory card array and measured the effect of the rise in temperature of the air on the heat transfer coefficient of the memory cards that were downstream in the air flow. Using these measurements, we devised a simple calculation model, called the thermal diffusion layer model, to calculate the heat transfer coefficient of multiple rows of memory cards. The rise in temperature of downstream memory cards due to higher temperature air can be evaluated with a parameter representing the delay of thermal mixing for air. The heat transfer coefficient calculated with the thermal diffusion layer model agreed with our experimental results.

Keywords: Forced Convection, Heat Transfer Performance, LSI Cooling, Memory Array

1. INTRODUCTION

The heat density of high-load electronics, such as high-performance servers and the inverters used in vehicles, has been increasing as a result of higher processing speeds and higher power outputs. Focusing on the heat density of high-performance server-CPU (Central Processing Unit) modules, the trend in development since 2000 has been away from bi-polar devices and toward low-power-consumption CMOS (Complementary Metal Oxide Semiconductor) devices. However, continued demand for higher performance has reversed the trend, and the heat density of CPU modules has begun to rise again. The method used to cool electronic devices depends on the heat generation level and the operating environment, whereas the available methods include natural convection cooling [1] , forced-air convection cooling, heat pipe cooling [2] , and liquid cooling. Forced-air convection cooling is the most often used because of its practical advantages in cost, installation, and maintenance. For these reasons, we have been developing a forced-air convection cooling structure to cool multiple LSI (Large Scale Integration) packages and memory arrays that are mounted on a flat wiring board in electronic equipment. A parallel flow method is used in which the LSI packages and arrays are cooled sequentially by air that flows parallel to the wiring board. A problem with this method is that the heat from the upstream devices raises the temperature of the cooling air so that its cooling capability is decreased remarkably for the downstream devices. Yanagida [3] analyzed the temperature distribution of air after heating devices and proposed a method that used the thermal diffusion layer model to calculate temperature distribution of LSI packages on a printed wiring board. The authors developed highly efficient cooling methods using inclined strip fins and air-jet cooling methods that were not affected by an increase in the temperature of the cooling air [4] [5] . Kheirandish et al. [6] analyzed forced convective laminar flow in a channel with a heated obstacle attached to one wall. Recently, Nakayama [7] reviewed various cooling methods of electronic devices. In this work we study forced-air convection cooling of multiple rows of memory cards placed on a printed wiring board. Each memory card has many DRAM-LSIs (Dynamic Random Access Memory LSIs) mounted on both sides (Figure 1). Some studies [8] on this topic reported on the cooling performance of a single memory card, but very few have dealt with rows of memory cards.

In this study, we made an experimental memory card apparatus with four rows in the flow direction and eight rows in the parallel direction. We measured the effect of a rise in the temperature of the air on the heat transfer coefficient of the downstream memory cards. Using these measurements, we then devised a simple model for calculating the heat transfer coefficient of multiple rows of memory cards. We compared the calculation and experimental results and found their values to be in good agreement.

2. EXPERIMENTAL APPARATUS

Figure 2 shows the experimental apparatus. The test section is placed in an acryl duct. The outside of the duct is thermally insulated with 50-mm-thick of glass wool. Cooling air from a blower flows into the test section after flow rate G and temperature T_{ain} are measured in chambers 1 and 2. Figure 3 shows the test section. Memory cards are four rows in the flow direction and eight rows in the parallel direction in the duct. Each memory card has five model-LSIs mounted on both sides. In the heat transfer experiment, any one of the four rows in the flow direction is replaced with heating model-LSIs and others are non-heating model-LSIs. The heating model-LSI is aluminum block with a stainless steel heater with heating rate Q of each model-LSI is 0.2 W. The surface temperature T_s of the heating model-LSI was measured with a T-type thermocouple, and the heat transfer coefficient h was obtained using Equation (1).

$$h = \frac{Q}{A_S\left(T_S - T_{ain}\right)} \tag{1}$$

Here, T_{ain} is the inlet air temperature, and A_s is the surface area. Also, the surface temperatures of non-heating model-LSIs were measured with thermocouples. In this paper, the number of rows of memory cards in the flow direction is expressed as N_{card} = 1 - 4 and number of rows of the memory LSIs in each card is expressed as N_{lsi} = 1 - 5.

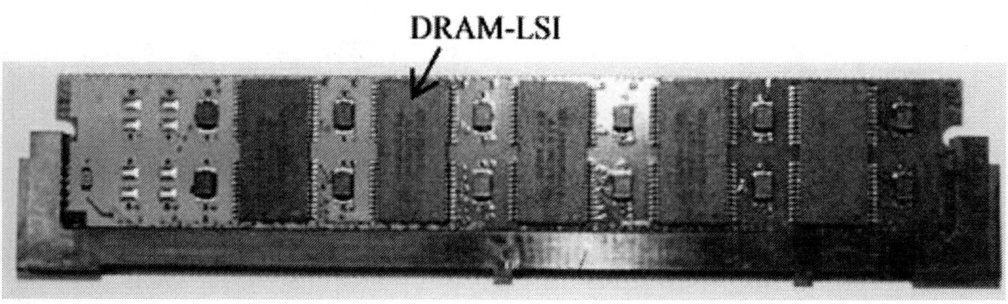

Figure 1. Memory card.

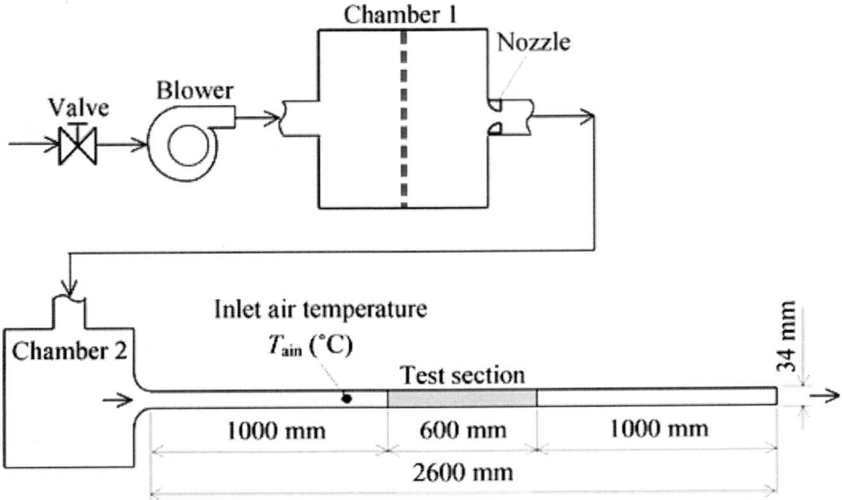

Figure 2. Experimental apparatus.

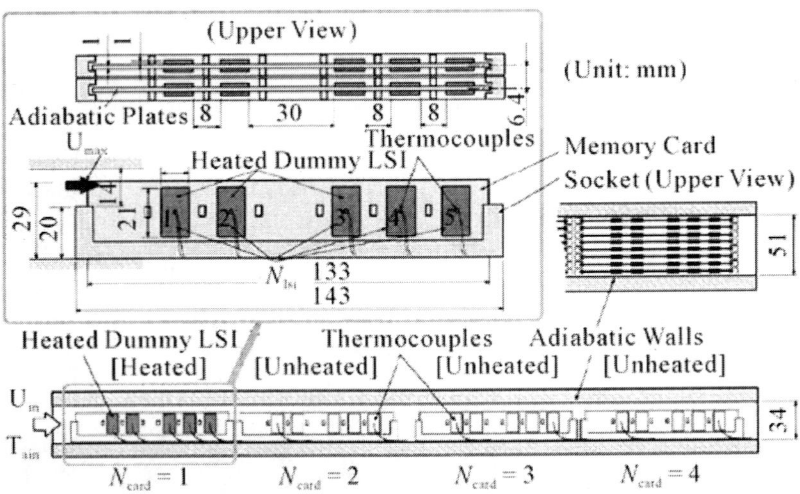

Figure 3. Test section of memory card array.

3. EXPERIMENTAL RESULTS OF TEMPERATURE RISE OF LSIS AND THERMAL DIFFUSION LAYER MODEL

First, we measured the temperature rise of non-heating LSIs when they were heated by the air flow from the upstream memory card. Figure 4 shows the experimental results of the temperature rise ΔT_a of the non-heating LSIs at the N_{lsi} = 1 or 5 position on the downstream memory cards at the N_{card} = 2 - 4 positions when the heating memory card was placed at the N_{card} = 1 - 3 positions. Figure 5 shows the parameter $\alpha_t = \Delta T_a/\Delta T_m$ which means the delay of thermal mixing in air, ΔT_a is the measured temperature rise of the non-heating LSI, and ΔT_m is the calculated temperature rise of completely mixed air. The temperature rise ΔT_m was calculated with the air flow rate and heating rate of the LSI. When the air is not mixed completely, α_t is larger than 1. The parameter α_t decreases downstream and comes close to 1 at all heating memory card positions, N_{card}.

Figure 6 shows the relation between the delay of thermal mixing in air α_t and the distance from the heat source X for various heating positions and flow velocities. We can obtain the relationship between the parameter α_t and the distance from the heat source X, similar to the thermal diffusion layer model proposed by Yanagida [3] .

Figure 7 shows the thermal diffusion layer model for our experimental apparatus. The dimensions of the heating LSI were as follows: height a = 1 mm, width b = 21 mm, and length L = 10 mm. The area of the thermal diffusion layer A is wider than the front area of the LSI (a ´ b) by the thermal diffusion thickness δ_t. Hence, we can write A = (a ´ b) + (2a + b) ´ δ_t. The delay in thermal mixing in the air flow $\alpha_t = \Delta T_a/\Delta T_m$ can be calculated with A_0/A, where A_0 is the cross-sectional area of the air duct. The thermal diffusion thickness δ_t is proportional

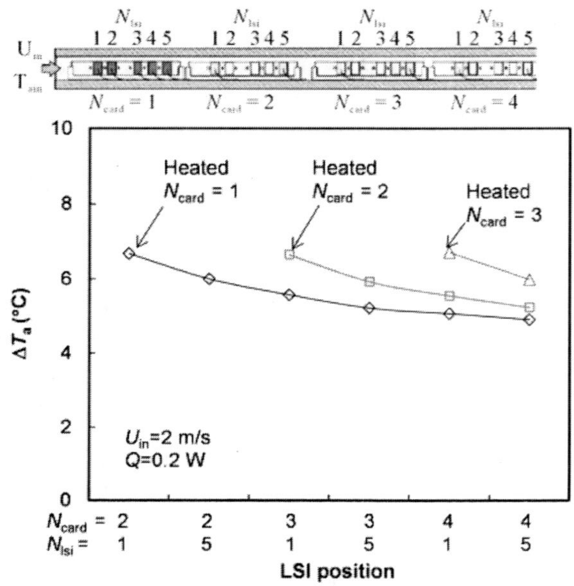

Figure 4. Temperature rise of LSIs.

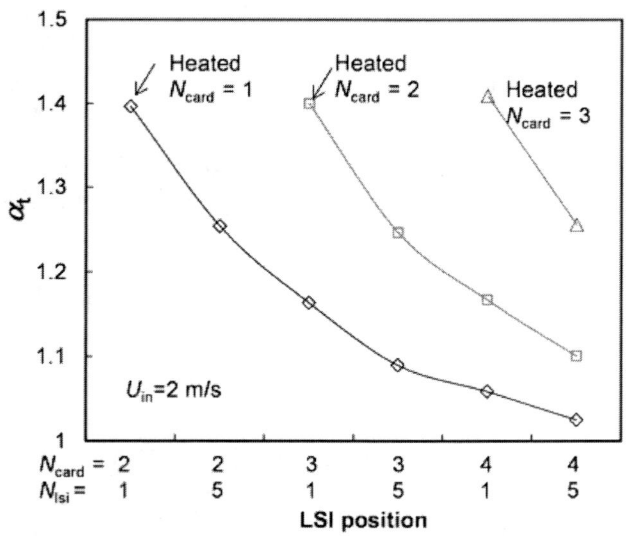

Figure 5. Delay of thermal mixing in air.

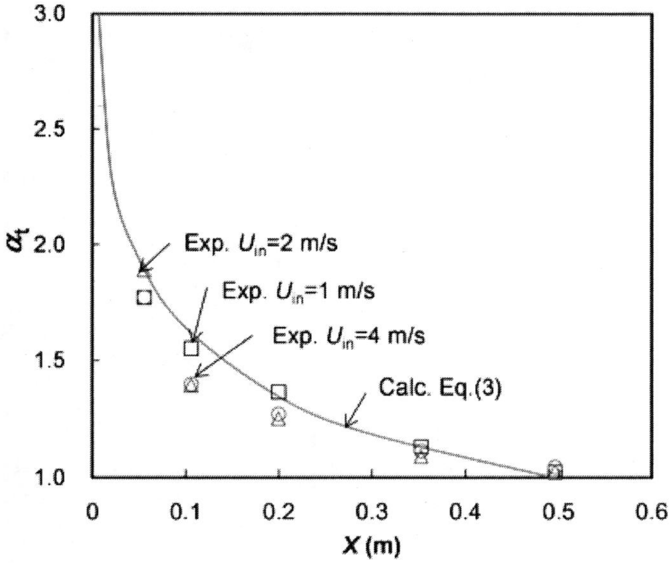

Figure 6. Relation between delay of thermal mixing in air and distance from heat source.

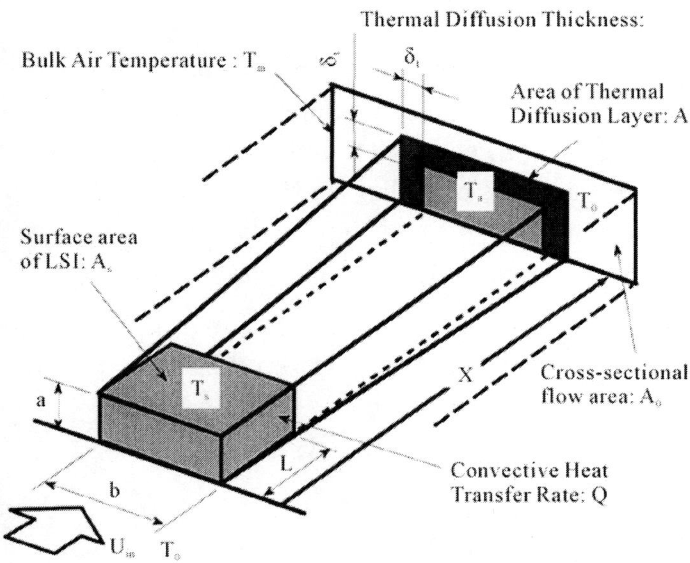

Figure 7. Thermal diffusion layer model.

to $\sqrt{X}$, and the proportionality constant is obtained using our experimental data shown in Figure 6, as follows:

$$\delta_t = 0.0032\sqrt{X} \tag{2}$$

The parameter α_t and the distance from the heat source X have following relationship using Equation (2):

$$\alpha_t = \frac{A_0}{A} = \frac{A_0}{ab + 0.0032(2a+b)\sqrt{X}} \tag{3}$$

Calculation results with Equation (3) agree with the experimental results in Figure 6.

4. EXPERIMENTAL RESULTS OF HEAT TRANSFER COEFFICIENT AND COMPARISON WITH CALCULATION USING THERMAL DIFFUSION LAYER MODEL

Figure 8 shows the experimental results of the heat transfer coefficient h of LSIs at the $N_{lsi} = 1 - 5$ positions on the memory card at the $N_{card} = 1$ position with a change in the inlet air velocity u_{in} when the heating memory card was placed at the $N_{card} = 1$ position. We see that the heat transfer coefficient h of the downstream LSIs is lower than that of the upstream LSIs, because of the temperature rise in the air flow.

The heat transfer coefficient h_S on a flat surface of a single LSI can be calculated with the following equations

[9] by using the Reynolds number, $\mathrm{Re} = \frac{u_{max} L}{\nu_a}$:

$$h_S = 0.664\,\mathrm{Re}^{0.5}\,\mathrm{Pr}_a^{0.33}\,\frac{\lambda_a}{L} \tag{4}$$

Here, u_{max} is the air velocity above the LSI's surface, L is the length of LSI, n_a is the dynamic viscosity of air, Pr_a is the Prandtl number of air, and l_a is thermal conductivity. The broken line in Figure 8 shows the result of calculation h_S using Equation (4), and it agrees with the experimental results of the $N_{lsi} = 1$ data.

Next, we calculated the heat transfer coefficient h_N coupled with the temperature rise in air using α_t. The heat transfer coefficient h_N can be calculated with the following equation:

$$h_N = \frac{1}{\dfrac{A_S}{GC_p}\sum_{i-1}^{N-1}\alpha_{t,i} + \dfrac{1}{h_S}} \tag{5}$$

Here G is the air flow rate, and C_p is the specific heat of air. The solid lines in Figure 8 show the results of the calculation h_N with Equation (5), and they agree with the experimental results of the $N_{lsi} = 2 - 5$ data.

Figure 9 and Figure 10 show the experimental results of the heat transfer coefficient h of the LSIs at the $N_{lsi} = 1 - 5$ positions on the memory cards at the $N_{card} = 2$ and 4 positions with a change in the inlet air velocity u_{in} when the heating memory card was placed at the $N_{card} = 2$ and 4 positions. The heat transfer coefficient h in Figure 9 and Figure 10 is almost same as that in Figure 8, and the calculations h_S, h_N using Equations (4) and (5) agree with the experimental results.

Figure 11 shows the experimental results of the overall heat transfer coefficient $h_{overall}$ of the LSIs at the $N_{lsi} = 1$ and 5 positions on the memory cards at the $N_{card} = 2$ and 4 positions when all memory cards at the $N_{card} = 1 - 4$ positions are heating. The results of the calculation with Equation (5) agree with the experimental results. Therefore,

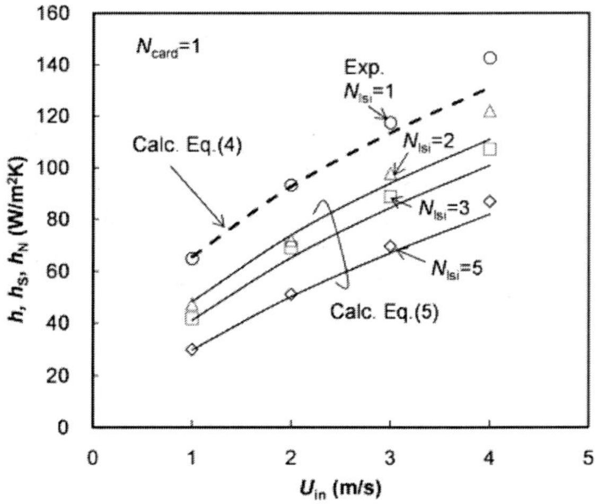

Figure 8. Heat transfer coefficient of LSIs N_{lsi} = 1 - 5 on memory card N_{card} = 1.

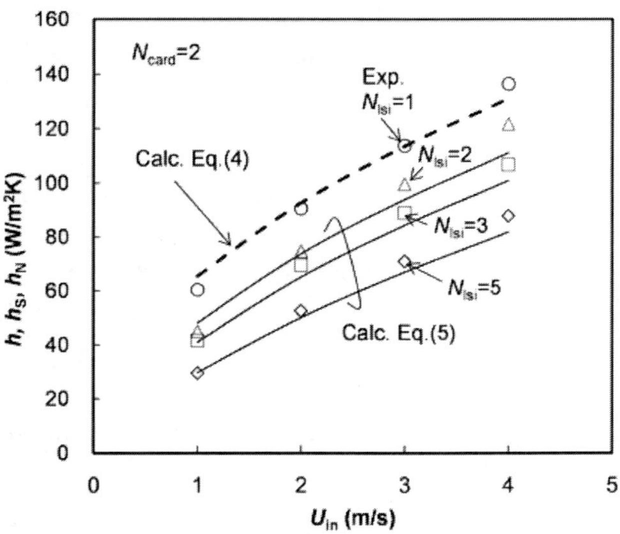

Figure 9. Heat transfer coefficient of LSIs N_{lsi} = 1 - 5 on memory card N_{card} = 2.

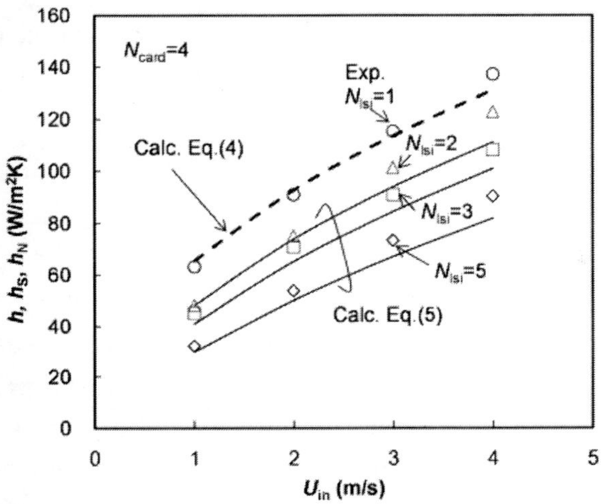

Figure 10. Heat transfer coefficient of LSIs N_{lsi} = 1 - 5 on memory card N_{card} = 4.

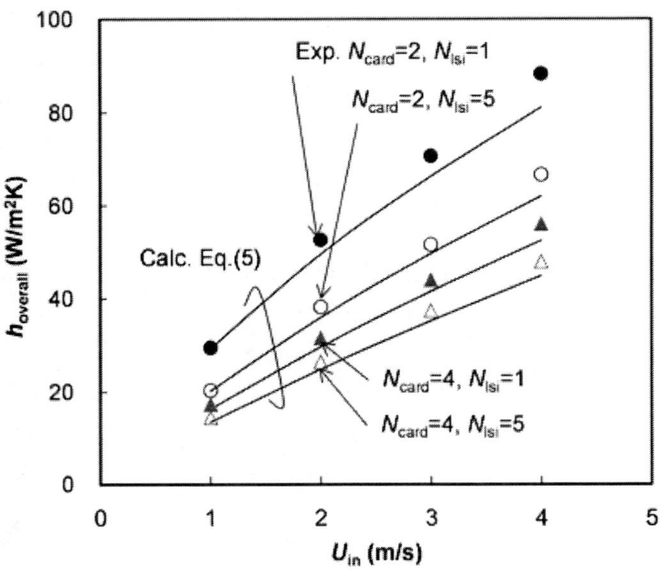

Figure 11. Overall heat transfer coefficient.

the thermal diffusion layer model of Equation (5) is a precise calculation of heat transfer coefficient of multiple rows of memory cards, despite that it is very simple. Our proposal of the thermal diffusion layer model of Equation (5) is thus very practical.

5. CONCLUSIONS

We made an experimental apparatus including a memory card array and measured the effect of a rise in the temperature of air on the heat transfer coefficient of memory cards at downstream positions in the air flow. Using these measurements, we proposed a simple calculation model of the heat transfer coefficient of multiple rows of memory cards. The following results were obtained.

1) The rise in temperature of the downstream LSI resulting from a temperature rise in the air heated by an upstream heating source can be evaluated with a parameter called the delay of thermal mixing in air, α_t. The parameter α_t decreases downstream and comes close to 1 at a large distance from the heat source X and α_t does not depend on the heating position or flow velocity.

2) The heat transfer coefficient coupled with the rise in temperature of the air through the parameter α_t can be calculated with Equation (5) and the thermal diffusion layer model. The calculation agreed with the experimental results. There results indicate that the thermal diffusion layer model is a simple and precise method of calculating heat transfer coefficient of multiple rows of memory cards.

REFERENCES

1. Nakazato, N., Hirasawa, S. and Mato, T. (1998) Natural Convection Cooling in Vertical Finned Plates in a Cabinet for Communication Equipment. IEICE Transactions on Electronics, 3, 421-426.
2. Toyoda, H. and Kondo, Y. (2013) Effect of Non-Condensable Gas Leakage on Long Term Cooling Performance of Loop Thermosyphon. Journal of Electronics Cooling and Thermal Control, 3, 131-135.
3. Yanagida, T. (1988) A Calculation Method for Temperature Distribution of IC Packages on a Printed Wiring Board. Transactions of JSME, Series B, 54, 1730-1735.
4. Atarashi, T., Hatada, T. and Daikoku, T. (1994) Forced Air Cooling Radiators for Electronic Equipment. Transactions of JSME, Series B, 60, 620-626.
5. Atarashi, T., Tanaka, T. and Tsubaki, S. (2012) Optimization of Heat Sinks with Plate Fins in Air-Jet Cooling. Proceedings of 3rd International Forum S
6. Kheirandish, Z., Nassab, S.A.G. and Vakilian, M. (2013) Second Law Analysis of Forced Convective Cooling in a Channel with a Heated Wall Mounted Obstacle. Journal of Electronics Cooling and Thermal Control, 3, 101-110.
7. Nakayama, W. (2014) Max Jakob Award Paper—Heat in Computers: Applied Heat Transfer in Information Technology. Transaction of ASME, Journal of Heat Transfer, 136, Article ID: 013001.
8. Yoneda, N., Arita, J. and Hosokawa, K. (1999) Thermal Analysis of Memory Module. Proceedings of JSME Thermal Engineering Conference, 31-32.

9. Holman, J.P. (1981) Heat Transfer (International Student Edition). McGraw Hill International Book Co.

INDEX

A

Air-Cooling, 252
analysis, 3, 16, 20, 27, 31, 32, 249, 257, 260, 281

B

Bio-Heat, 66, 85

C

Condensation, 35, 36, 49, 60, 61, 62, 63
Copper, 235, 236, 238, 239
counterflow, 2, 8, 26

D

diffusion bonding, 1, 2, 3

E

entropy generation, 143, 144, 145, 146, 147, 149, 152, 181, 184, 185
exchanger (PCHE), 1, 2, 3, 9, 10, 11, 24
exergy loss, 211, 212, 213, 214, 217, 218, 219, 220, 221, 223, 229
Experiment, 252
experimental, 2, 3, 12, 13, 20, 301, 302, 305

F

fan coil unit, 212, 213
finned heat sink, 113, 114, , 134, 137
Forced Convection, 60, 294

friction coefficient, 26, 45, 144, 156, 157, 165, 277

H

Heat Sink, 126, 252, 253, 267, 272
Heat transfer, 22, 28, 30, 36, 291, 303, 304
Human Eye, 65, 66, 84, 85, 86

L

Laminar flow, 276

M

Memory Array, 294
micro photo-etching, 1, 2, 10
microchannel, 1, 2, 3, 9, 10, 11, 57, 58, 59, 155, 186, 247
Microchannel, 1, 8, 10, 29, 36, 60, 61
microgrooves, 235, 236, 237, 239, 240, 241, 242, 243, 245
micro-pipe, 144, 145, 148, 175, 176, 178, 179, 180, 181

N

Numerical Analysis, 252, 256
Numerical simulation, 36
Numerical Simulations, 66

P

printed circuit heat, 1, 2, 3, 9, 10, 11, 24, 30, 31, 32

R

radiant floor, 212, 231
radiation, 70, 71, 74, 79, 114, 255, 256, 259, 260, 266

S

Simulation, 62, 85, 154, 276

T

Thermal behavior, 236, 249